Mining Economics and Strategy

Ian C. Runge

Published by the
Society for Mining, Metallurgy, and Exploration, Inc.

Society for Mining, Metallurgy, and Exploration, Inc. (SME)
12999 East Adam Aircraft Circle
Englewood, Colorado 80112
(303) 948-4200 / (800) 763-3132
www.smenet.org

SME advances the worldwide mining and minerals community through information exchange and professional development. With members in more than 70 countries, SME is the world's largest association of mining and minerals professionals.

ISBN 10: 0-87335-165-7
ISBN 13: 978-0-87335-165-2
Ebook: 978-0-87335-291-8

Library of Congress Cataloging-in-Publication Data

Runge, Ian Charles, 1952–
Mining economics and strategy / Ian Charles Runge.
316 p. 21.6 × 27.9 cm.
Includes bibliographical references and index.
ISBN 0-87335-165-7 (softcover)
1. Mining engineering--Cost effectiveness. 2. Mineral industries--Government policy. I. Title.
TN274.R78 1998
338.2'3--dc21

98-24661
CIP

Contents

List of Figures

List of Tables

Preface

Mining used to be a business primarily focused on the technical aspects of getting valuable ore out of the ground and extracting the minerals in a metallurgically efficient way. Although the importance of these skills cannot be denied, a narrow focus on technical issues is no longer sufficient to guarantee success, even in the richest orebodies.

Skill in economics is an essential partner to technical skill in every step of the mining process. The economic way of thinking starts from before the first drillhole is put in the ground. It includes not just the most economic way of mining, but also the most economic way of going about assessing mining projects. It directs mining strategy and takes equal notice of the forces of world progress and the forces governing individual human action.

The scope of this book includes what is meant by a cost-effective mining scheme. It includes the economics of information, as well as the procedures for rational evaluation of mining projects under uncertainty. It reexamines the definition of ore from an economic perspective. In particular, it specifically considers the economic influence of scheduling on ore reserves.

This book addresses discounted cash flow (DCF) techniques—the most widely used evaluation technique for investment decision making—in detail. Although this technique has been known and used in the mineral industry for decades, the widespread use of spreadsheets has been a feature of DCF evaluations only since the mid-1980s. The assumption of the use of spreadsheets is a significant point of differentiation in this text from previous mining-focused economics texts. It means that more meaningful examples can be included. Formulas developed to overcome previous computational difficulties have been omitted. Further, examples in the text are available in spreadsheet format.* The

* A CD-ROM containing spreadsheet files and sample financial modeling software is available from the author to original purchasers of this book (see instructions at the end of the book). If instructions for obtaining this CD-ROM are missing from the last page of the book, please contact the author directly (E-mail: irunge@runge.com) or via Runge Ltd., www.runge.com.

application of DCF techniques in an operating mine environment is given expanded coverage, and examples are drawn from real-life studies.

The differences between economic decision making—a forward-looking task—and the reporting of results via accounting methods—a historical or backward-looking activity—are reviewed. Nevertheless, it is not the intent in this book to provide a comprehensive coverage of general economics or corporate finance principles. (The book is intended as a stand-alone text in mining economics and strategy; however, for corporate finance issues of a generalized nature, a text such as Brealey and Myers [2003] is highly recommended.) This book gives extensive coverage to capital and to decision-making procedures associated with capital investments in a risk environment. Comprehensive case studies for capital investment in an operating mine are included.

Many traditional approaches to mine valuation overlook important strategic elements, leading to results that frequently fall short of expectations. If, for instance, one mine plan can accommodate change more easily than another plan, but at some cost, how can the value of that flexibility be understood? Many of these elements are definable in advance; the difficulty is in finding the mechanism to incorporate them into decisions. The theory of decision making under uncertainty is briefly examined, and the applicability of this theory to understanding the risk/return trade-off is highlighted.

Comprehensive examples investigate value from a risk reduction perspective and from a perspective of expected return on investment. A case study using probabilistic analysis derives analytically tractable results for valuing equity participation in a major mining project under conditions of uncertain offtake.

A theme in the book is that many mining projects that fail to achieve expectations do so because of their inability to adapt to change. This problem can be partially addressed through greater predictability in future conditions. It can also be addressed through mining schemes that can sustain returns over a greater range of foreseeable future conditions.

In the context of making investments, assets can be differentiated into components reflecting their contribution to profitability, adaptability, and risk reduction. This book sets out a new technique allowing calculation of capital that is at risk from capital that is not at risk. The use of this technique is a precursor to mine design that is less sensitive to changes that are outside mine operators' control. The technique promises significant advance in the way that investments are made and capital is valued in the industry.

Although the book starts from and largely maintains a technical perspective, it also recognizes that the institutional environment within which the industry operates has a significant influence on the degree of success of mining ventures.

The book finishes with an overview of mining strategy, with a strong emphasis on knowledge effects. It suggests that, in mining at least, imperfections in knowledge play a significant role in determining mining strategy and are also a significant contributor to less-than-perfect decision making evidenced throughout the mining world. It draws upon current trends in strategy in the wider business environment, applying them to the mining industry. It sets out some promising future directions for mining strategy and potential for added value through enhanced decision making in the industry.

Acknowledgments

This book is the outcome of technical and economic analysis spanning 30 years of involvement in mining projects, major investments, and proposed investments worldwide. The outline for the text was initially prepared in 1984 for a series of courses entitled "Design of Integrated Open Pit Coal Mines" that I conducted while I was employed with Runge Ltd. (Brisbane, Australia). Since 1992, the economics components of this coursework have been refined and expanded through two courses: (1) "Mining Economics," used as the basis of the first 11 chapters of the text, and (2) "Capital Investment Strategy," used in the balance of the text. These courses have been presented throughout Australia, North America, Africa, and Southeast Asia by myself and by Runge Ltd. personnel.

I am indebted to the hundreds of attendees at these courses for substantial feedback and suggestions—and in particular to Mike Rowlands, Hugh Thompson, Tony Savva, Donna Luxton, Christian Larsen, Tony Kinnane, and other senior staff from Runge Ltd. for contributions and critiques of many parts of the work. The subject matter is a vast one that cannot be comprehensively covered through the examples and experiences of one person alone. I acknowledge the inevitable errors of omission, as well as the equally inevitable errors of commission, that must remain my own. Feedback and further contributions toward the advancement of mining economics along the lines developed in this text are welcome and appreciated.

CHAPTER 1 Introduction

The economics of the resources industry are unique. All mining is subject to uncertainties not applicable to other industries. Every mine is different. Industry economics are difficult to quantify and categorize. Information is very costly.

In major mining countries, there is now a real dichotomy. The products of the minerals industry are essential primary ingredients in almost everything used in an advanced society, yet their availability is often taken for granted. In the developed world, the value of mining is increasingly being called into question. The difficulty in making profits is compounded by political uncertainties and environmental restrictions on top of the uncertainties created by nature.

Against this backdrop, however, actual production in many developed countries has *increased.* Despite declining prices, as well as profitability that frequently falls short of expectations, more capital continues to be injected into the industry. Many of the factors that lead to profits or losses escape recognition if conventional tools of analysis are used.

Low profitability may be the order of the day for many participants in the industry, but this is not universal. In some cases, the finding of a rich orebody, either by skillful exploration or by chance, has been the key to success. Yet there are many examples of rich orebodies not producing profitable mines and of mediocre orebodies turning into successful long-term enterprises. In some cases, the ability of a company to anticipate the market has resulted in profits above expectations. More commonly, an unexpected market downturn has resulted in losses despite efficiency elsewhere in the mining operation. Even in this environment, superficially similar mines have turned in vastly different profit performances. Management decision making and *factors other than market pricing and orebody characteristics* probably have a much greater influence on industry profitability than has been acknowledged up until now.

One of the primary tasks of this book is to examine some of the characteristics that lead to increased value but that escape analysis on more superficial grounds. A basic premise of the book is that a lot more can be done to maintain consistency and predictability in the economics of mining. Price unpredictability and orebody characteristics have in many cases been inappropriately used as excuses for underperformance. Even where these unpredictable elements are real, the opportunity for mine design that is less sensitive to such change has frequently been overlooked.

Two decades ago, *technical* skills were the primary ingredient for efficient mining management. With the internationalization of world industry, this has changed. Technical skills are no longer sufficient. The underlying premise of this book is that *every technical decision must consider and be guided by the economic consequences.* The criterion upon which the whole enterprise is judged (namely, shareholder value) should be the same criterion used to judge each day-to-day or medium-term decision. The more consistency there is between the criteria applied by the different personnel making these judgments, the more likely it is that the company objective will be achieved.

This book looks at mining economics through two lenses. The first, akin to the traditional economic view, takes the economic environment as neutral. This view is most appropriate to operating decisions at mine sites. In such an environment, many economic elements are outside the decision maker's control. Capital has already been sunk. The product selling price and many other factors are taken as given. Many alternatives are excluded because of the limited time to implement them. These sorts of issues are addressed in the first part of the book (Chapters 2 to 11).

The second lens focuses on a broader range of issues. It recognizes that whether a mine is performing or not performing is unlikely to be due to some *technical* cause alone—though some technical cause might be held to blame. It recognizes that choices are not played out in a vacuum; instead, they are subject to changes in the world at large over which the decision maker has little control. A new mining development justified on a *rising* metal price will yield a return dramatically below expectations if the price trend turns out to be *falling.* If the trend was conceptually predictable, then a *strategic* decision that overlooked such a possibility is the primary source of the subsequent underperformance. Many of these strategic decisions are considered unimportant at the time they are made.

Although in this book there is a clear demarcation between the first style of economic analysis and the second, in reality the distinction is not so clear. Most elements that lead to success from a strategic viewpoint cannot be divorced from the usual economic decision making that is undertaken every day in an operating mine. If mine operators are not clear about the strategic direction the mine is following, then the problem is only partly addressed.

Most approaches to economics assume that whatever one person or one company does, the whole market will be unchanged. An extra ton of coal produced will not change the world price of coal. This is the traditional (nonstrategic) approach and is the usual assumption in the first part of the book. The strategic view is somewhat different. It acknowledges that the outcome for each participant depends upon the choices (strategies) of all the other participants. The extra ton that *one mine* makes will not affect the world price, but if the changed economic forces that allow *this mine* to produce an extra ton also apply to its competitors, then surely the competitors will increase output as well. The world price *will* change.

Strategic decision making has been subjected to a lot of analysis in the economics profession through game theory. Unfortunately, very little of this has reached the everyday business world. Some of the tools for undertaking these evaluations are described in this text—not so much for their path-breaking characteristics, but primarily to demonstrate the applicability of already-well-understood techniques to mining activities.

The text has been prepared with the following three key guidelines in mind:

- It focuses on the primary economic evaluation techniques in use in most mining applications. Older, little used techniques that have been superseded by spreadsheets and computer models either have not been included or have been mentioned only briefly.
- The text aims to *apply* economic evaluation techniques, not just explain them. In addition to subjecting techniques to some scrutiny, the examples provide a template for making economically sound operational, investment, and planning decisions with confidence.
- It provides indications of how reliable or unreliable certain techniques are in the real-world environment (i.e., a world of change, risk, and uncertainty). It offers techniques and mechanisms to adapt the standard methods of project analysis and decision making so that they reflect what is happening in the changing world of today.

In an ideal world, an economic evaluation will yield the same result no matter who is undertaking it or who is making the decision. In the real world, however, any evaluation is based on inputs that are quite uncertain and objectives that are commonly ill defined. This uncertainty may have a greater influence on the decision making than any analytically derived (expected) return. The cost of resolving the uncertainty (e.g., the cost of extra drilling) can make the project uneconomic.

In such an uncertain environment, the narrowly defined but commonly used economic evaluation techniques frequently lead to unreliable answers. This does not mean that they should not be used, however. Indeed, the reverse is the case. Uncertainty demands *increased* rigor and even more systematic processes in evaluation. The value is frequently in the process more than in the result:

- The process imposes a disciplined framework for all important information upon which to base the decision. Even if the final decision has to be made on individual perceptions of risk and uncertainty, application of the techniques set out in this book will cause at least someone on the project team to examine each primary input in some detail.
- The process provides a benchmark for tracking performance after implementation. A base case (the "expected" plan) also defines the importance, or relative unimportance, of each element in the plan. The knowledge as to

whether something is important or unimportant is essential for setting management priorities and for highlighting "day 1" deviations that would not otherwise be noticed.

- New techniques that *can* model the influences of uncertainty are becoming available and useable. Uncertainty does not necessarily mean "risk," and mine designs that are capable of producing as planned *despite* uncertainty are more robust than designs that are more constrained to a fixed set of conditions.* To understand the robustness of a particular mine plan requires an effort an order of magnitude beyond that for developing a simple base case plan. However, a standard economic analysis is still the starting point for any such extended effort.

Chapter 2 provides general examples of how economic evaluation techniques are applied at various stages of the mine evaluation process; it also introduces the basic concepts of economic evaluation. Chapter 3 defines the key elements and objectives of mine planning and mining operations in economic terms. It provides guidelines for the amount of detail and the relative importance of the various components in each stage of the evaluation.

This book aims to help practitioners make better decisions by using economic analysis tools applied to mining applications. This is a forward-looking approach and contrasts with traditional accounting, which is primarily a historical or backward-looking activity. Chapter 4 sets out the differences between "costs" as applied in this economic sense and "costs" as commonly applied in other applications. This chapter also includes expanded definitions and examples of fixed costs, variable costs, opportunity costs, and marginal costs and revenues.

Almost all decision making in a mine involves trading off current costs and benefits with expected future costs and benefits; hence, almost all mine evaluation must entail comparisons involving the time value of money. Chapter 5 sets out the methodology for this form of evaluation and provides examples of the use of discounted cash flow (DCF) analyses for typical mine decisions.

Costing and evaluation of any mining development are necessarily based on a specific mine plan, which has to be prepared assuming certain orebody characteristics. However, orebodies are seldom clearly defined, and the effort to find and delineate them is itself an economically significant task. The economics of mining will determine what parts are or are not included in the definition of *ore* (i.e., ore in an economic sense, not in terms of some statutory definition). When mine economics change, the amount of material in the ground does not change, but the amount of *economically viable* ore does change. The amount of economically viable ore is also dependent on the assumptions used for its calculation and can change with a change in

* In some of the finance and economics literature (see, for example, Knight [1921]), the term *risk* refers to inputs or outcomes to which a probability distribution can be applied. In this literature, the term *uncertainty* is used in cases where probability distributions cannot be applied. This book adopts the lay person's concept of risk, akin to "exposure to the chance of injury or loss." More detailed analysis of these terms, as well as the impact of uncertainty and risk on decision making, is scrutinized particularly in Chapter 12.

assumptions. For example, a starting assumption of small-scale mining implies high unit costs, and a mine evaluated based on this cost structure may indeed indicate limited reserves consistent with this mining method. Alternatively, a starting assumption of bulk mining—with its attendant low costs—allows deeper and lower-grade parts of the deposit to be mined economically and therefore included in the economic reserve. With no change in physical characteristics, large *economic* reserves may be demonstrated for the same orebody consistent with this bulk mining method. Chapter 6 addresses these issues and provides the economic tools for evaluating parts of the deposit and deciding whether those parts should be included in the plan or not.

The most useful and easy-to-use application of time-value concepts is to translate into comparable terms the capital costs of equipment and alternatives that have different mixes of capital and operating cost. This process is addressed in Chapter 7.

Mines that have been operating for some time all have records of the costs of operating the equipment, and this is an important input for any economic evaluation. Nevertheless, this information is reliable only for the equipment in use and for the conditions applying up to that time. Chapter 8 builds on the capital cost estimates from Chapter 7 and sets out the procedures for estimating the operating costs of mining equipment. The estimating techniques are derived from first principles and apply to equipment working in any conditions; this allows auditing and extrapolation of existing cost information, as well as comparison with potential new mine equipment.

With all the important tools having been set out in the preceding chapters, Chapter 9 describes how to use these tools in both simple and more advanced ways to make investment decisions. Chapter 10 presents a comprehensive case study of a typical investment decision in an operating mine.

Chapter 11 is aimed at new mines and steps through the detailed buildup of all of the inputs for a complete cash flow.

The examination of mining *strategy* begins in Chapter 12, commencing with an introduction to the capital investment process and the risk/return dichotomy. This chapter also examines the nature of capital choices and some of the theoretical foundations for rational choice under uncertainty. It discusses what is different about mining-type investments compared to choices in consumer markets or investment decisions in the equities market.

Mining enterprises are complex industrial organizations, and decision making has to involve shortcuts. Over time, shortcuts that prove reliable become embedded in the culture of the organization—continuing to be used even when the economic environment has changed. Some of these short-form evaluation techniques are scrutinized in Chapter 13. This chapter also looks at some of the more common models of business strategy and their applicability in a mining environment.

Chapter 14 extends the decision criteria introduced in Chapters 5 and 9 to make explicit the allowances for uncertainty. Two studies are presented that apply the theoretical models introduced in Chapter 12 to machine selection and to mine equity valuation.

In Chapter 4, the economic concept of "cost" is introduced, and a differentiation is made between the cost of a *decision or commitment* to perform some event and the cost of the *event*. The difference is evident when the objects of choice (alternative ways to develop a mine, for instance) are themselves paths into the future that contain follow-on choices and options to change. Common evaluation techniques focus on comparing events that assume this path into the future is irrevocable. Chapter 15 explicitly separates out the cost of a decision from the cost of the event. It sets out specific tools for differentiating capital that is at risk from capital that is not at risk. The tools permit ready comparison of projects with different risk/return profiles. Among projects subject to change, the technique demonstrates substantial differences that remain unrecognized under the use of conventional evaluation methods.

With the advance of computing technology and economic science, mining projects are increasingly subject to analytical scrutiny. Yet the value of mining companies remains strongly influenced by human factors not capable of analysis by even the most sophisticated tools. Chapter 16 puts this issue into a historical and a knowledge-oriented perspective; it suggests a number of areas where value may be added to mining enterprises substantially beyond the limits of the tools set out in the previous 15 chapters.

The book also includes a glossary of economic terms as they apply to mining and an appendix with a set of tables of financial data.

CHAPTER 2

Integrating Economics Into Mining

Even relatively small mines are complex business undertakings, and in running these businesses only a small number of decisions warrant the time or cost of a comprehensive economic analysis. Furthermore, most choices do not *require* such an analysis. Tools for making economically based choices in any complex business span the full spectrum from simple rules of thumb to elaborate and comprehensive financial models. This chapter provides general examples of how economic evaluation techniques are applied at various stages of the mine evaluation process.

MINE-PLANNING PROCESS

A complete task of planning and operating a mine involves at least three components:

- a technical component
- a narrowly focused economic component
- a more broadly based economic component, including financial and business elements that influence mine performance within industry at large

The technical component concerns mine layout, equipment productivities, alternative production schedules, and mine operating requirements. These latter requirements include, for example, the explosive usage per year, the number of persons required, and the fuel usage per machine per operating hour. Usually this work focuses on technical criteria only. This component of a mine plan will be unchanged whether the project is economical or not. The technical component defines all of the important elements concerning the implementability of the proposal. Some of these elements may be economically important and some quite unimportant as far as their impact on mine economics. For example, two completely different mining schemes may have very similar operating and capital costs.

The narrowly focused economic component applies operating and capital *costs* to the technical schedules. It analyzes alternative schedules and alternative equipment in economic terms—e.g., the price per ton. It also builds up and examines unit costs, such as the fuel cost per liter, annual fuel cost for the whole mine, and labor cost per person per year. The object of this phase of the work is to allow comparison of options in economic terms.

Often these narrowly focused economic evaluations concern just some *component* of a project—for example, even if the mine itself is uneconomical, a valid economic analysis of the cheapest way to move *waste* can still be undertaken. Most mine evaluation is limited to technical work, coupled with this narrowly focused economic phase.

The more broadly based economic, financial, and business component aims to understand the degree of viability of a plan and how the plan fits within a wider corporate context. Whole-project viability is a function of what other projects the company may have available, as well as what other companies (suppliers, customers, and competitors) are doing. This phase of evaluation also examines the relative risk associated with investment decisions and the sensitivity of the plan to factors outside management control. It foreshadows likely difficulties in implementation. It attempts to position the mine for likely change. From a business perspective, this broadly based economic analysis is vital, since market valuations—and with them, the stability of the corporate structure—are integrally bound with management's ability to *deliver* what the mine plan *indicates* is achievable.

Mining is an expensive activity, the cost-effectiveness of which demands these economic analyses. Mine *evaluation* is also an expensive activity and demands its own economic analysis. Is another $500,000 expenditure for exploration likely to yield as much increase in shareholder value as a $500,000 expenditure on metallurgical testing? The task of the mine planner is to develop a mine plan that will maximize shareholder value, and one of the components in this process is the cost of, and cost savings flowing from, the mine-planning task itself. Mining economics is also concerned with the proportion of resources that should most appropriately be directed to this task and how these resources themselves, once allocated, should most appropriately be directed.

APPLICATION OF ECONOMIC EVALUATION PROCEDURES

Cost estimating and economic evaluation are integral parts of the planning process. Since mine planning is an iterative process, the form of economic assessment changes as the precision and reliability change with each successive iteration. Evaluation work itself has a cost. Undertaking evaluations in an economic way demands that alternatives that are unlikely to find their way to subsequent development be eliminated early in the evaluation cycle. Expenditure that subsequently proves to have no enduring value is minimized, and the same resources put into evaluation yield higher-value results. The examples set out in the following sections illustrate the application of economic evaluation procedures during some sample phases of a complete mine-planning study.

Justifying Exploration Expenditure

The evaluation of every potential mining development starts somewhere. As a minimum, some estimate of how much ore is in the ground must be available, coupled with some estimates of the costs of extracting it. Yet even with the simplest understanding of these numbers, it is possible to eliminate many alternatives early on by formulation of minimum guidelines. Exploration expenditure itself is prioritized on the basis of the deposits or potential deposits that, if found, are likely to yield the greatest value per dollar of expenditure.

Grassroots Exploration In grassroots exploration, an initial starting point for evaluation is an *assumed* orebody. Prior to any substantial expenditure, the following question must be asked: If this exploration is successful, what is the *best* deposit that can be expected? The hypothetical deposit can be described in terms of tonnage, grades, depth, and physical location.

An economic analysis should be conducted on this hypothetical deposit. Of course, given the paucity of data, this economic analysis may take only a few hours of work. Yet valuable insights are possible. Clearly, if the evaluation of the best hypothetical deposit does not indicate viability, then lesser deposits will not either. Sensitivity analysis will also yield important information. The economics of a hypothetical gold deposit, for example, might suggest that ore running 3 g/t at 100 m depth has about the same economic value as ore running 5 g/t at 200 m depth. This knowledge is a valuable guideline for exploration. Many companies have expended millions of dollars on exploration, followed by subsequent evaluation, only then to find that the project is not viable. Many projects could have been proven nonviable before *any* expenditure.

If an idealized orebody is not viable, exploration might still be undertaken—but clearly some alternative justification for the exploration is necessary. This early economic evaluation is designed to focus early attention on the primary objective—i.e., creating a profitable mine.

Ongoing Exploration Ongoing exploration around some known orebody serves two purposes: (1) finding more reserves and (2) understanding existing reserves to ensure greater reliability. Additional reserves usually translate into additional value in a straightforward way. Unfortunately, reserves that are simply understood better are not necessarily worth more than reserves that are understood to a lesser degree.

For coal-mining applications, ongoing exploration seldom consists of finding more reserves; more likely, it is concerned with understanding and proving up already-delineated reserves to a greater degree of reliability. The raw volume of coal in the ground may be unimportant compared to the degree of faulting, folding, and quality changes that occur with an already-delineated reserve. How can one tell whether extra reserves are economically a more important target for exploration than are faults, folds, or quality characteristics? Justifying this ongoing exploration is a complex task. More reserves usually mean higher potential production rates, efficiencies due to larger equipment, and fixed costs spread over larger outputs. Usually these gains are easy to quantify and this style of exploration is readily justifiable, but ongoing exploration aimed at improved reliability in the estimates yields its benefit in terms of lesser risk. Risk-based gains are harder to quantify. Some tools for valuing risk reduction are set out in the second part of this book.

For many metalliferous mining applications, ongoing exploration to delineate mining reserves can often be more easily quantified, since drilling costs are a significant contributor to the mining cost structure. A surface drillhole to intersect a deep underground orebody might cost 20 times as much as a drillhole underground from a position adjacent to the orebody, but it may not yield any more information. At some point the cost of drilling itself starts to impact mine economics. This problem is rarely an issue in coal mining.

Example 2.1:

Assume the costs of delineating a block of ore today are \$5.00/t, with mine economics as follows:

costs excluding exploration	\$30.00/t
average exploration/delineation cost	\$5.00/t
average return, or selling price	\$40.00/t
profit	\$5.00/t

If the company requires a 15% return on investment, then how far into the future can it afford to block out mining reserves?

Answer:

Expenditure of \$5.00/t now to delineate reserves that will not be mined until next year has to be treated in the same way as any other investment that yields its benefits in the future. If costs are independent of reserves, then reserves should be delineated only when they are needed for mine planning—any work delineating reserves in advance will reduce profits. How far in the future can we afford to do this?

In this case, if reserves could be delineated as a direct operating expense immediately prior to mining, then each ton of ore would yield \$5.00 profit. If, for mine-planning purposes, we have to delineate reserves 1 year prior to mining, then how much profit will we make? The cost is again \$5.00/t to delineate the reserves, but if this money were invested elsewhere for the year, it would return \$5.75 (\$5.00 × 1.15)—so the decision to delineate reserves in advance must account for this lost opportunity to earn \$0.75 with the funds. The effective "profit" reduces to \$4.25/t.

Five years into the future, the \$5.00 expenditure now equates to "cost" then of \$10.05 ($\$5.00 \times 1.15^5$), or a lost opportunity to make \$5.05 of profit elsewhere. Drilling out reserves 5 years in advance, when the benefit is received at that later time, reduces the effective profit to nil.

In the preceding example, operating costs were assumed independent of reserves, and the reserve delineation was limited by the present value of future benefits. The solution given is only a partial answer to the problem, and not strictly the correct answer, since it implies that reserves should be delineated to the extent of profitability. This is not necessarily the case. If costs *per ton* reduce with larger reserves—and this is the normal case, with costs of access ramps and driveages spread over larger tonnages—then an optimum reserve definition in terms of years of production is possible independent of profitability. Such a case is described in Chapter 6.

This simple example and the more extensive examples in Chapter 6 explain why the world has reserves of bauxite exceeding 200 years at current production rates, whereas the reserves of silver amount to only about 20 years at current rates (e.g., see Crowson [1991]). It does not mean that the world is about to run out of silver or that silver mining is less economic; it means only that it is uneconomic to prove up the reserves now when there is no potential to exploit them until a long time in the future. Indeed, some of the world's greatest precious metals mines, in production for 50 years or more, have rarely had more than 5 years of proven reserves.

In many forms of mining, the cost of proving reserves is so high in proportion to the overall cost of production that it is more economical to proceed initially with a higher level of uncertainty. Choices for the long term—such as shaft, access, and infrastructure development—assume a long life even though reserves are reliably understood for only a few years. The alternative—to prove up 10 or more years of reserves first—subtracts from the realized return more than it adds by way of reduced risk as reflected in the lower cost of capital.

Strategic Assessment of Mining Projects

Once a potential mining development has been identified, how does technical evaluation proceed in a rational way, accounting for the economics of information as well as the economics of mining?

A strategic assessment—based on economic criteria—is an important element in this rational decision-making process. This assessment, as well as how well it is understood, is probably the greatest factor differentiating successful projects and successful companies from those that are less successful. The difficulty is that, from very early in the evaluation process, some broadly assumed final development scenario defines a path along which new information and new studies are directed. If information were free and took no time to prepare, all information could be found and a simultaneous comparison of alternatives undertaken. However, because information is gathered and evaluated sequentially in successive rounds of evaluation, the process is path dependent. Alternatives that might be equally attractive in economics terms may never get to be compared. One of them may be eliminated early in the evaluation process when its good characteristics are less understood but when the bad characteristics of some other alternative are not yet evident.

The following two examples illustrate the importance of incorporating economic criteria early in the strategic assessment of mining projects.

Example 2.2:

An open pit coal-mining project has up to 100 m of overburden that is soft enough for digging without blasting. Bucket-wheel excavators are ideal tools for this digging environment. Yet just because the ground can be physically dug using bucket-wheel technology does not mean that this is the most economical mining method. An inefficient alternative method may be cheaper than an idealized bucket-wheel method. Without an early strategic assessment, there is a risk that subsequent exploration is directed in such a way that the initially chosen alternative becomes a self-fulfilling and self-justifying choice. The path dependency in this process applies to personnel as well as information. If you select a team for the technical evaluation whose expertise is in bucket-wheel technology (and why wouldn't you, if this is the mining method that appears to be best?), then this team has an inherent disposition toward this technology.

Example 2.3:

In the late 1980s, a large international mining company, whose expertise was almost solely in underground mining, planned the development of a new underground precious metals mine in an African country. More than three-quarters of the reserves were too deep for open pit mining. A large (and quite profitable) underground mine was planned. Just prior to commitment, personnel not previously associated with the project undertook a fresh review (a last look!) of open cut potential. Dramatic changes followed. Although reserves as stated favored underground mining, these reserves had been delineated under the assumption of underground mining costs. There were substantial additional shallow reserves that were quite profitable to mine assuming open pit mining costs. Moreover much of the "waste" in the open pit mine—material that had to be hauled out of the pit anyway—contained precious metal for which the marginal returns from treatment substantially exceeded the extra cost if the material was otherwise dumped as waste. The mine commenced development in 1991 as a large and quite profitable open pit precious metals mine. Further, the development of the underground mine was not precluded since the reserve depletion associated with the open pit mine involved only minimal impact on the economic viability of any underground development.

Strategic assessment must be undertaken early in the evaluation of a project, and it must be as broadly based as possible. The direction and priorities for subsequent exploration and evaluation effort may change radically depending on which constraints of the intended mining method need to be satisfied. In addition, the skill requirements of the evaluation and development team may be quite different for each alternative. The presumption of one or another method will result in path-dependent subsequent decision making. This path-dependent influence introduces the risk that choices will favor the skills of the participants rather than (more correctly) being made based on the inherent characteristics of the deposit itself.

Guidelines for strategic assessment of mining projects are introduced in Chapter 3.

Selection of Reserve Blocks and Phased Mine Development

Optimum mine development typically requires a balance between

- averaging the good and bad characteristics of a resource to balance equipment utilization and for consistency of mine output
- deliberately selecting more profitable blocks in the early mine life to maximize the advantages of early cash flows

The averaging process might mean blending for consistent (average) quality. It might mean mining blocks of high waste:ore ratio simultaneously with blocks of low waste:ore ratio. The focus on selecting the more profitable blocks early is to balance the higher profitability of these blocks with interest repayment and capital imposts that are most severe at the start of the mine. Clearly, before an optimum mine layout (size, shape, etc.) and mining sequence can be determined, the relative economic value of each block of ore (or coal) must be understood. The following example demonstrates the types of issues that arise concerning mining sequence; this phase of the mining project assessment is addressed in Chapter 6.

Example 2.4:

An open pit mine has some very attractive reserves in one pit area, but these can be accessed only by first excavating a deep boxcut. The ore from the initial excavation returns \$50/t but costs an average of \$80/t to mine. Once the initial excavation has been developed, succeeding mine blocks average only \$30/t to mine against the same \$50 return. Opening up the area quickly will result in a high initial cost in boxcut development, but it will return faster cash flows once succeeding blocks are mined. Opening up the boxcut slowly will allow profitable mining elsewhere in the lease to "subsidize" the initial development of this pit, and the subsidy will eventually be recovered from the profits of the succeeding blocks. What is the optimum development timing and production rate from this pit?

Comparison of Mining Equipment Options

Mining operations continually change. Existing equipment becomes less suitable to the task. New equipment is continually developed to move mining quantities more cheaply. In an ongoing mining operation, or in the detailed planning phase of a new mining development, the most common task requiring economic analysis is the exercise to compare two or more potential alternatives.

Example 2.5:

You need access to a difficult mining area for a period of about 8 months. You can access it via an existing ramp, or you can spend \$50,000 of development expenditure to install a new access. The new access will improve equipment productivity by 20%. In either case, you have enough equipment on-site to move the required tonnage. Is it economic to install the new access, or should you persevere with the existing access for 8 months?

Answer:

In this case, probably all expenditures are operating expenditures, so simple comparisons without tax, without capital cost considerations, and without time-value discounting are appropriate. Equipment productivities and costs do not have to be accurately known to achieve a reliable estimate of the change in productivity or cost.

Example 2.6:

A spoil pile has failed and has covered over half of the width of coal over a 120-m-long section of pit. You have mined around it, and now have to decide whether you should abandon the coal or use your existing equipment on overtime to clear the failed spoil to recover the coal. You have already uncovered the coal once, so uncovering it again is going to be expensive—but the waste now overlying it is a lot less than for any other coal in your mine. You have sufficient inventory of coal in the pit and sufficient main waste removal capacity to allow you to maintain production regardless. Should you abandon the coal or not?

Answer:

The fact that the coal has already been uncovered once is irrelevant. If all alternatives have the same revenue and the same fixed costs, then the comparison is based on operating costs. The present value of the operating costs of a mine plan including re-uncovering the coal has to be compared with the present value of the operating costs of a mine plan excluding re-uncovering the coal.

Examples 2.5 and 2.6 describe typical cases encountered in day-to-day operations. Similar comparisons relating to two or more potential alternatives are encountered in longer-term planning.

Example 2.7:

Additional stripping capability is required in an open pit mine. Alternatives vary from use of contractors (the alternative with the lowest capital cost but highest operating cost), to large rope shovels and trucks (the alternative with the lowest operating cost but highest capital cost). The alternatives also vary in terms of annual amount of waste moved. In addition, alternatives have varying degrees of risk and also differ in the constraints and flexibility in the pit operations.

There is no one technique suited to all of these forms of evaluation. Nevertheless the following guidelines should be used:

1. If there is a different capital and operating cost mix between options, then a discounted cash flow is essential—including consideration of taxation effects. If the production rate between alternatives is the same, then *usually* the comparison can be undertaken in isolation with the rest of the mine. Chapter 5 describes the standard discounted cash flow analysis for this type of evaluation.

2. If the production rate varies from one alternative to the next, then the comparison must consider other influences. Normally, the whole mine must be incorporated in the economic comparison—but considered separately from the equipment comparison task.* For an unchanging selling price, fixed costs of infrastructure spread over a larger mine output will favor larger-tonnage options.
3. Where the comparison is based on incremental costs over a short term (as in Example 2.6), then only direct *cash* costs should be considered. In this kind of example, depreciation, interest charges, or "fixed" operating costs (such as head office overheads) may be ignored.

For most of the cases involving comparison of alternatives, some form of discounted cash flow technique is appropriate and yields reliable results. The discounted average cost technique (a variant of the standard discounted cash flow model) is set out in Chapter 5.

Overall Project Evaluation

The evaluation of an overall project is usually undertaken by using discounted cash flow techniques. A technical study initially yields tabulations of the required equipment, personnel, and supplies necessary during each year of preproduction and production to produce a certain amount of mine output. The costs of purchasing and operating the equipment are combined with personnel and other costs to determine the overall cost structure of the mine. The expected revenue from the sale of the mine output is also estimated. A cash flow tabulation sets out all of these cash inflows and outflows year by year (including tax payments) to determine the annual cash flows for the project.

In most mining projects, there is a large cash outflow in the early years due to purchases of mine equipment, followed by cash inflows later in the mine life. After discounting of future cash flows back to their equivalent present values, the overall net present value (NPV) of the project is determined. If the owners consider that there is sufficient value in proceeding with the project compared to any alternative courses of action, then they can make a decision accordingly.

Every company has limited resources, and (unless the project is small compared to the resources of the company) the decision to proceed will be dependent on concurrent similar evaluations of alternative projects. For this reason, whole-project analysis is a time-consuming procedure. In the end, support from company boards will be forthcoming only if the funding for a given project fits into a time frame consistent with all other demands on the company's

* This style of analysis is very common—and very commonly misunderstood. The issue of optimum mine output must be considered separately from the issue of the mine output that flows from one style of development or another. A mine constrained by waste removal may show very high returns from additional waste removal capacity—this is an argument favoring additional capacity, but it is *not* an argument favoring any particular type of overburden equipment. Once the "optimum" rate of additional waste capacity is established, then the secondary issue—deciding which is the most appropriate equipment to use—should be based on the lowest discounted average *unit* cost of waste removal, not on the whole-mine economics.

resources. The following example demonstrates some of the issues that may arise in evaluations of alternative projects.

Example 2.8:

A new project requires initial capital expenditures of $100 million, followed by annual after-tax profits of $20 million per year over 10 years. The return on investment is 15%. The company has an alternative but higher-risk investment proposition requiring $50 million of capital; this project yields 18%, but if it proceeds the company will have insufficient funds to proceed with the larger project. The company is also concerned because the smaller but notionally more profitable project uses a lot of diesel-powered equipment; if there is a major increase in the price of fuel oil, then the return on investment will be greatly disadvantaged. Which project should proceed? In this example, project uncertainty has to be presented in probabilistic terms to assess the risk. Higher risk in itself is not a criterion for rejecting the second project unless one project exceeds the threshold criterion of risk and the other project does not. "Risk" itself is a function of the size of the company. The evaluation must also consider the return on investment on the difference in capital between the two alternatives.

The preparation of cash flow tabulations for whole-project analysis is introduced briefly in Chapter 5 and covered comprehensively in Chapter 11. Where alternative investments are subject to different amounts of risk and sensitivity to change, more sophisticated analysis is usually necessary, some elements of which are covered in Chapters 14 and 15.

CHAPTER 3

Cost-Effective Mining Schemes

What constitutes a cost-effective mining scheme? Once two or more alternatives are available, selection of a cost-effective mining scheme implies some objective measure of comparison. Simple rules such as minimum operating cost or maximum internal rate of return often prove inadequate for comparisons of alternatives with different capital requirements or different risks. The first part of this chapter examines some of the key elements that characterize different mining schemes from this perspective.

However, there is an even more fundamental element in the quest for a cost-effective mining scheme. This element is concerned with how the alternatives chosen for evaluation come to be considered in the first place. A systematic method of evaluation also requires a mechanism to *discover* these possible alternatives. This systematic process of discovery, coupled with refinement of rules, is addressed in the second part of this chapter. The final part of the chapter looks at the various phases of evaluation and how the reliability, return on investment, and risk change with increasing refinement of a mine plan.

KEY ELEMENTS

For initial evaluation, the key elements of any cost-effective mining scheme can be categorized into four groupings:

- operating cost characteristics
- capital requirements and capital characteristics
- sensitivity to change and scope to adapt to change
- consistency with knowledge and philosophy of the owner

Operating Cost Characteristics

Low operating costs are always a desirable objective. If a given company has lower operating costs than its competitors, then for the same mine output selling at the same market price, the company will have larger cash flows. This gives the company greater protection against variability in market price, and it means that the company can continue in production at prices that its competitors cannot match. The company may be incurring losses, but the competitors will be losing more money.

An objective of low operating costs is particularly important for export-oriented mines, since the price of mineral commodities sold on the export market is usually more volatile than that of similar commodities supplied to domestic markets. Indeed, some domestic contracts use "cost plus" pricing, in which case the incentive might be to *increase* operating costs.

Low operating costs can come about because the deposit is better (shallower, higher grade) than competitive deposits. Apart from a corporate objective to seek out such deposits, this characteristic is largely outside the control of the mine planner. Low operating costs can also come about through more capital-intensive mining schemes, but this route has a price in terms of higher capital costs or perhaps less flexibility. In such cases the choice is far from clear-cut.

Capital Requirements and Capital Characteristics

Issues relating to capital fall into three groupings: capital-intensiveness, timing of capital expenditures, and the amount of capital.

Capital-Intensiveness The term *capital-intensive* is used to describe investments for which there is additional expenditure of resources now with the expectation of lower costs later. It implies that, in operation, fixed costs make up a large component of total costs. When practitioners talk about implementing a mine plan that is more capital-intensive than some alternative plan, they are usually referring to the expenditure of (more) capital at the start of a mine to achieve low or lower operating costs throughout the mine life. Sometimes additional capital is expended to achieve a reduced risk or to set up the mine for easier expandability. Capital intensification is a sought-after objective in most large-scale mining in the world, but schemes that are more capital-intensive are not necessarily desirable, nor are less capital-intensive schemes necessarily undesirable. A higher capital cost may mean a disproportionate exposure (financial risk) in unstable political environments. Capital-intensive schemes are frequently less flexible than alternatives, and this may result in *higher* technical risk or less ability to change with changes in the market.

Delayed Capital Expenditure If all other things are equal, delayed capital expenditure is a desirable objective, since capital incurred later has a lower cost (in present value terms) than the same expenditure incurred today. Nevertheless, all other things may not be equal. Higher operating costs incurred by delaying capital might mean that a company is foregoing a return that may exceed the return they are getting on their money elsewhere. If properly incorporated into the mine plan, delayed capital commonly allows three other advantages:

1. The characteristics of the mine will be better understood at the time of commitment. Less capital may be required.
2. Delayed capital is subject to less uncertainty, implying a lower cost than if all of the capital were expended initially, when knowledge of the mining conditions is less well understood.
3. Even if the total capital is unchanged in present value terms, it can be partly funded from cash flow. Exposure—the maximum amount of cash outflow needed before cash starts flowing back from the investment—is reduced, and financing costs may be lower.

Capital Expenditure Appropriate to the Structure of the Company Whenever a new proposition for investment is advanced, one of the first questions asked is, How much will it cost? The implication is that capital *costs* should be minimized. This implication is unfortunate. Although the owners of companies are sensitive to any proposition for capital expenditure, the raison d'être of an "investment" is to "use" capital. Efficient deployment of a firm's resources (including capital) is certainly important, but minimization of capital per se is *not* a key element in the selection of a cost-effective mining method.

What *is* important is a total capital requirement consistent with the size of the company. A medium-sized mining company may be able to afford $200 million of investment comfortably from existing resources (retained earnings, established lines of credit), but beyond this it might have to raise additional equity or use more costly forms of finance. The *extra* cost of this *extra* capital might be very high—particularly if it puts at risk other established businesses of the company. The marginal cost of capital as a guideline for investment is discussed at length in Chapter 14.

Sensitivity to Change and Scope to Adapt to Change

The future never materializes according to any one plan or set of plans, yet shareholder value is directly related to how consistently businesses perform, year in and year out, despite changes occurring around them. This robustness in the face of change may be an inherent characteristic of a deposit, but more often it is a function of (1) the mining method and equipment chosen, (2) whether this equipment or method is adaptable or not, and (3) what institutional constraints are placed on adaptation.

Sensitivity to Influences Outside of the Company's Direct Control In most mining applications, the key factors impacting mine economics that are outside the direct control of the company are the market price and volume, exchange rates, and mineral deposit characteristics. If waste hardness is unknown, for example, a mining scheme for which the productivity and costs are relatively insensitive to this factor is preferred. The cost of truck/shovel schemes, for example, is not as sensitive to waste hardness as is the cost of bucket-wheel excavator schemes. It may be preferable to adopt an alternative, less sensitive scheme even if this scheme has a higher cost. The cost of removing the uncertainty must be balanced against the higher cost of the less sensitive scheme.

At the start-up of a mine, many deposit characteristics may be quite unknown—favoring *initial* mining schemes that are less sensitive to these characteristics (even at higher costs, initially). Once the mine has been operating for some time, these characteristics may become better understood, and the bounds of predictability may increase. Mining schemes that *are* sensitive to these characteristics can be implemented if they offer cost advantages within these bounds. The mine plan should also recognize the need for later change.

Scope for Production Variation The biggest unknown in many mining projects is the sale price and offtake of the company's products. For this reason, almost all new mines commence production at annual rates below their theoretical optimum rate. If the mine were to commence production at a higher rate initially, some or all of the production might have to be sold into less-developed markets with a low marginal return. If production is increased as markets expand or as older mines close, consistency in market pricing can be maintained. A more cost-effective mine plan is one that includes within its scope these production increases (and perhaps production decreases toward the end of mine life) in line with market requirements.

Economic Limits Unconstrained Within the economic life of a mine, enormous changes always occur in the efficiency of mining and transport and in the price of the mine's product. The establishment costs and disestablishment costs are very substantial, and frequently the most profitable period occurs after the mine is more than 10 years old. At this time the mine might even be exploiting reserves that were considered to be quite uneconomic at the start. Although doing so may be difficult to quantify on a discounted cash flow basis, mines should be designed wherever possible to allow this ease of mine extension. Major crushing facilities should be placed above mining areas that have the lowest probability of ever being mined. Alternatively, the present value of cost savings for closer placement of surface facilities should be clearly documented, so that when change occurs in the future, operators *then* will be alert to the economic logic that was originally employed; this reasoning can be overruled if it proves to be no longer relevant.

Consistency With Knowledge and Philosophy of the Owner

Each year large numbers of mining properties change hands, and under new ownership there are frequently quite dramatic changes in mine performance. Equipment productivity calculations sourced from statistics or manufacturers' handbooks frequently overlook the fact that a machine produces some "rated" output only under the guidance of a skilled operator. The skills of this operator, the instructions for the task, and the ease with which other mine operations mesh with the task at hand all play a role in achieving expected outputs. What differentiates mining from almost all other industrial activity (factory processes and the like) is that the mine changes daily as reserves are exploited. The shape and characteristics of the next day's reserves might differ greatly from past experience, and the current smooth-running production operation cannot necessarily be translated automatically into a smooth-running operation tomorrow.

This changed production environment is mirrored in the changed decision environment, and the mine's time-honored rules for supervisors and managers may be inapplicable with changes in the mine plan. Many decision makers may not even be aware of this fact. The systematic processes set out in the balance of this chapter are designed to forestall this problem, and some of the short-form evaluation and decision techniques are subject to additional scrutiny in Chapter 13.

Simplified Operation and Reduced Complexity Mines that have been in operation for some time develop a very sophisticated corporate culture—a set of written and unwritten rules that evolve over time. It is these rules and procedures that allow the efficient transmission of knowledge throughout the workforce. Because it evolves over time, this tacit knowledge is commonly quite unrecognized.

The value of this knowledge becomes abruptly apparent when a team of "trained" personnel engages in a technically similar task that is *not* similar in economic terms. The early stages of a mine represent an archetypal situation. In this case the less-than-smooth interactions among untrained personnel (or personnel trained under an inappropriate set of rules) can impact production efficiency in dramatic ways. Production from large manufacturing plants, as well as from most capital-intensive enterprises, is quite consistent from day to day, but *all* mines change from one day to the next. It is very difficult to establish enough consistency for new personnel to understand the operation well enough to operate efficiently.

New projects should be designed to allow time for these tacit rules to evolve; during this initial learning period, a guideline of "the simpler the better" should be the order of the day.

Mining Method That Reflects Corporate Philosophy Just as day-to-day operational work is subject to tacit knowledge that is seldom consciously understood, so too is decision making at a higher level in the corporate structure.

Example 3.1:

In the late 1970s and early 1980s, many of the world's large oil companies entered the coal-mining business seeing it as an extension of their existing "energy" businesses. However, by the mid-1990s, almost all of these companies had sold out of their coal interests after failing to assimilate them into their larger corporate structures. Although the two businesses supply markets with many similarities, the production side of the mining business—where most difficult decisions are made—is quite different from that of the oil business, where production is relatively less important. The relative sizes of the businesses, financing and accounting inconsistencies, and the less fungible nature of coal compared to oil were just three other substantial differences that made this transition difficult.

In a marketplace subject to change, success that depends on decision-making consistency goes to two kinds of organizations. In the first kind, new developments should be consistent with the philosophy and capital structure of the organization. Alternatively, the organization itself must enjoy a corporate culture that can assimilate rapid change. In the absence of these conditions, new developments must be approached with caution. Mining schemes that allow sufficient time for adaptation (in the kinds of decisions being asked of senior management) should take preference over superficially more efficient methods that demand more sophisticated, higher-level decision support.

Comparing alternatives based on simplistic economics (e.g., lowest capital cost, lowest operating cost) is no longer sufficient. In a changing world environment, the key to maintaining and improving the economics of mining lies in schemes that are adaptable enough to accommodate a wide range of circumstances and for which cost-efficiency can be sustained over this range of circumstances. The economic trade-off must incorporate average and marginal costs, as well as uncertainty and risk, and must be understandable by production and operations personnel to be a useful guideline for all of their decision making.

THE SYSTEMATIC PLANNING PROCESS

Once a proposition to use a particular mining system has been advanced, the mine design must correctly reflect the economics of this type of scheme.

Example 3.2:

In open pit coal mining, dragline mining schemes require the mine to be planned out in long, narrow strips. The so-called technical explanation of this constraint is that a dragline can transport the waste only a limited distance. However, on examination this is not a technical constraint at all. Draglines can transport material very long distances by rehandling material—multiple numbers of times if necessary. The constraint is actually an economic one. The cost of dragline waste removal increases dramatically once material must be moved a distance exceeding twice the operating radius of the machine. A small amount of rehandling is permissible, but once a dragline scheme requires more extensive rehandling, other mining methods are usually more economic.

For loader/truck methods, different constraints apply. Unlike with the dragline, actual transport distance or the shape of the mine is seldom a limiting factor; however, the efficiency of using trucks is very sensitive to the change in elevation through which the load must be carried. The cost of truck haulage increases substantially as the open pit is deepened. This usually constrains the mine layout to match haul distances to the required change in elevation. It also places other constraints on the positioning of dump stations and out-of-pit waste dumping areas, and it governs the pit shape for optimum placement of in-pit refill.

If there is no mine plan yet, then the costs of mining or relative costs of alternative mining schemes are unknown, so how can the mine be planned out to reflect these costs? The answer to this dilemma is that the mine-planning process is an iterative one. The expected structure of the mining costs for any one type of mine design is derived from a previous, less detailed plan of the mine or a similar mine. The first broad-brush plans of the mine and each successive iteration of the planning cycle derive and refine the rules of evaluation for subsequent phases in the cycle. Original rules derive from established corporate experience and from generalized rules of thumb such as described in this text. The systematic planning process is set out in a three-dimensional flowchart in Figure 3.1. The process consists of a similar series of steps in each phase, each undertaken in the same order. Three such phases are illustrated in Figure 3.1; however, in practice there may be any number of phases. The same series of steps are undertaken in varying amounts of detail, depending on the precision, economic action, or decision being sought.

The kinds of decisions and studies through the various phases of the planning cycle are set out in Table 3.1.

Figure 3.2 focuses on the evaluation task. This task applies for every phase of the process. The evaluation task has a rule-related element and an action element. The rule-related element does the following:

1. Confirms that the rules of evaluation preselected for this phase are consistent with the findings from the phase. For example, if alternatives are to be primarily compared on the basis of their cost of production, the rule-related element confirms that the cost of production is a primary factor that differentiates one alternative from another.
2. Establishes, for the alternative that has not been eliminated, a refined set of rules (guidelines for the next phase). This refined set of rules is the starting point for comparison among subsets of the noneliminated alternative in the subsequent phase of evaluation.

The courses of action open after any evaluation are

1. Implement the proposal.
2. Examine the proposal in more detail. Proceed to the next phase of study, seeking more reliability in the estimates prior to making a decision.
3. Examine more alternatives—either at this same level of precision or by stepping back to a more broad-brush level.
4. Abandon the project.

ECONOMIC DATA AT EACH PHASE OF THE PLANNING CYCLE

Progressing from one phase of evaluation to the next is not simply a matter of increased precision in the result—each phase also eliminates from contention alternatives that might point the mine in some entirely different direction. For this reason, the rules for evaluation during the initial broad-brush and strategic planning stages must be subject to scrutiny at the highest levels. It is at this stage when the greatest number of alternatives are available and it is easiest to incorrectly eliminate entire classes of potential plans.

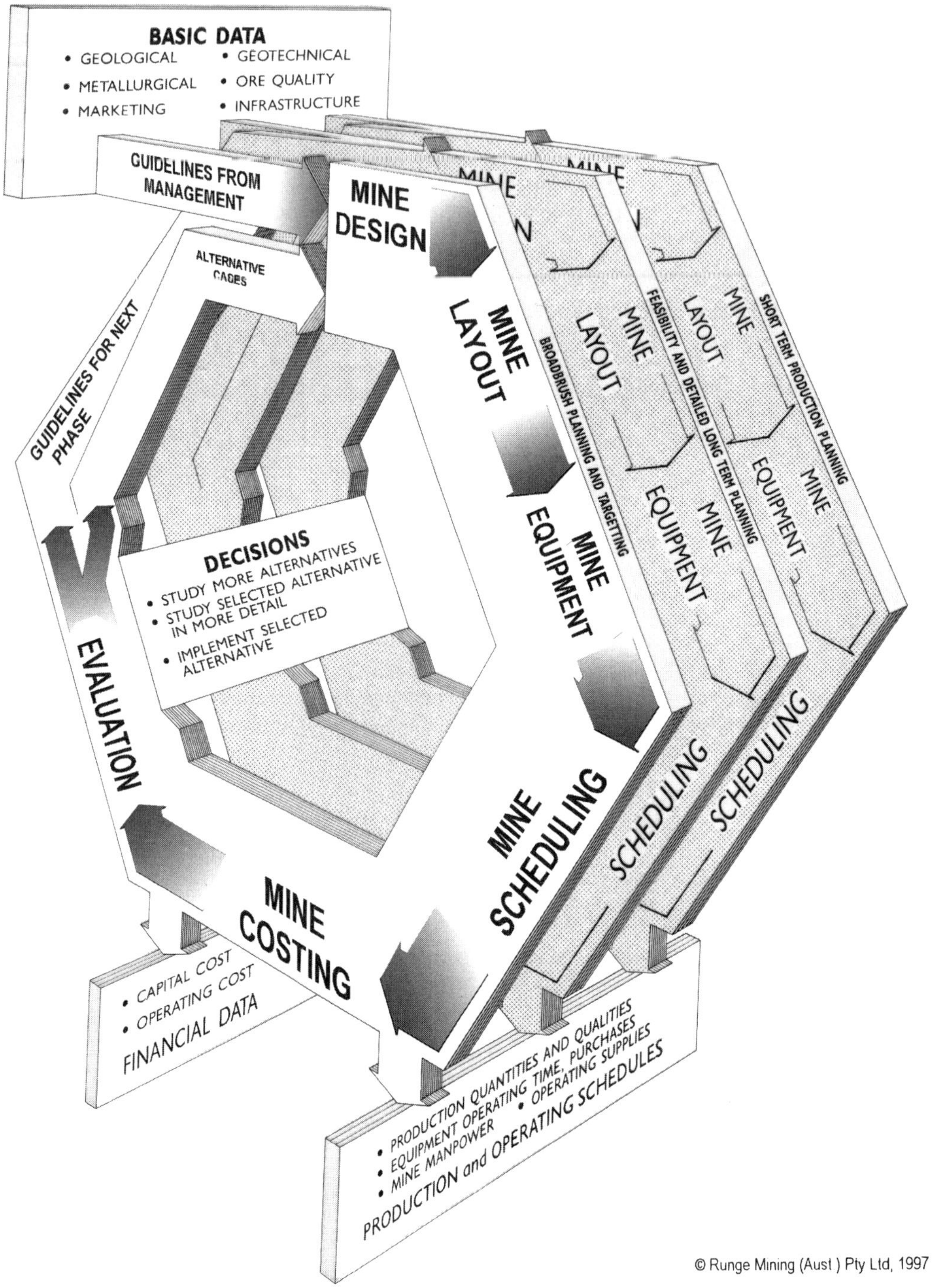

Source: Figure has been derived from work undertaken by the author at Runge Ltd., and permission to include this diagram from this source is acknowledged. Runge (2000) outlines a more comprehensive generalized model of decision making under uncertainty, of which this figure represents a narrow mining-related subset.

FIGURE 3.1 Iterative mine-planning flowchart

TABLE 3.1 Phases of the planning cycle

Phase	Economic Action, Decision, or Study
First review of a new project (a broad-brush mine plan)	Use rules of thumb based on industry experience to ascertain the likely project economics. The fundamental economics of most mines are set in place by no more than four or five factors, which usually are depth, waste:ore ratio, ore characteristics (type, grade, plant yield), the selling price, and the distance to the market. Identify any particular production, market, or other constraints or risks that could potentially impact the project revenue.
Strategic planning and development strategy	This is the key phase of any project. It is the phase when critical decisions are made concerning development strategy and alternatives with substantial differences in economics are eliminated. Following this phase the fundamental economics of the mining operation are largely fixed. Changes in strategic direction have major influences on costs, risks, and capacity to accommodate change. Once a strategy has been adopted, subsequent work (the phases following this step) results in more evolutionary influences on costs and risks. Evolutionary change may not result in large changes in costs of production, but it may still result in significant change in profitability. Further, the cumulative effect of extensive evolutionary change over time may permit change in broad strategy, with substantial impact on costs of production. The scope for such cumulative evolutionary change is also a part of the strategic planning phase of evaluation.
Detailed long-term mine plan or feasibility study	This phase involves at least two different types of economic evaluation: **1.** Within each cost center or definable activity, the optimum mining scheme has to be determined (e.g., for waste removal, the proportions of waste moved by trucks, cast blasting, or contractor). This type of economic evaluation is undertaken by discounted average cost calculation or similar techniques restricting inputs to those factors likely to influence the result. **2.** For whole-mine evaluation, the complete operating and capital costs for the mine must be developed and tabulated—leading to the whole-mine cash flow analysis.
Equipment selection	Optimization of equipment sizing and trade-offs between different machines must be undertaken on a relative cost basis—usually using some form of discounted cash flow analysis. Frequently, equipment selection also involves the comparison of mining alternatives whereby some equipment purchases are delayed (either to save initial capital or because there is insufficient working room available in the early stages of mine life). These evaluations involve alternative mine schedules.
Mine development phase	In the mine development phase, most of the decisions influencing the economics are complete. This phase involves setting up the systems to compare actual costs with planned costs and to quickly highlight anomalies. The process also involves auditing of accounting procedures to ensure that precedents established for tracking of costs do not result in "day 1" deviations that are then not discovered in subsequent accounting within the mine.
Yearly planning	This is normally undertaken by using costs already established from prior experience.
Monthly planning	This is normally undertaken by using costs already established from prior experience.
Weekly and daily planning	Guidelines prepared from more exhaustive general studies should be summarized to allow field personnel to make day-to-day decisions based on economic criteria.

Broad-Brush Planning Phase

The first phase of evaluation, broad-brush planning, is aimed at providing the guidelines for exploration. It is not uncommon for hundreds of potential mining scenarios to be examined at this stage–and appropriately so, since the cost of examining hundreds of alternatives at any greater level of precision may be prohibitive.

It is customary to speak of "precision" or "reliability in cost estimates" in terms of being within some percentage range of the actual value. For broad-brush analyses, the reliability of the result could easily be correct to within ±20%, but this figure is misleading because it would be true only under the assumption

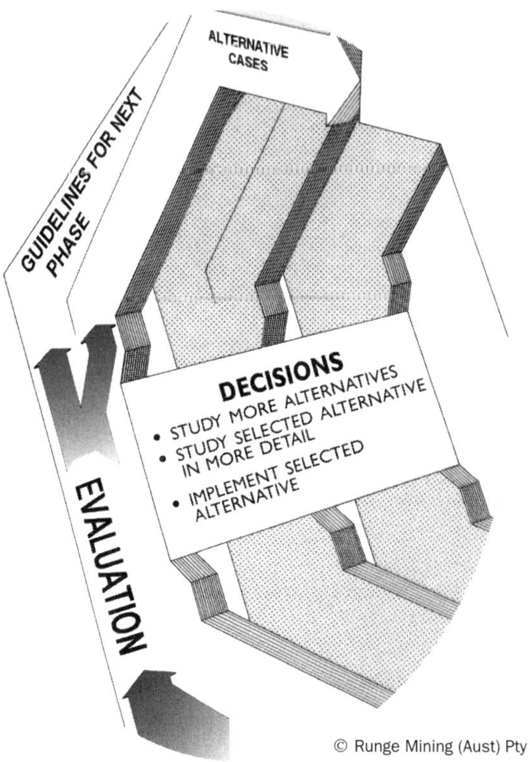

FIGURE 3.2 Decision alternatives in the planning process

that the mine plan can in fact be implemented. The validity of this assumption is demonstrated only after the detailed feasibility study.

Nevertheless, for comparing alternatives, broad-brush plans provide reliable guidelines for the likely economic characteristics of the mine—assuming such plans can be implemented. Assuming no systematic bias or omissions, a whole-mine cost estimate may be within ±20% even if the individual inputs are accurate to only ±40%. Individual estimates that are high can balance the estimates that subsequently turn out to be low.

Strategic Planning Phase

The second phase of the planning cycle, the strategic planning phase, fills a critical role in optimizing the mine. It sets in place the fundamental economic structure for the entire mine. The priority in this phase of the planning cycle is to ensure three things:

1. Understanding of what characterizes this (potential) mine compared to other mines or what characterizes this mine's products as judged by the customer—and adoption of a strategy that maximizes this value. Even if the fundamental cost structure is not different from that of other mines, there may be significant points of differentiation from other mines. Differentiation provides the basis for premium pricing of products, sales of products

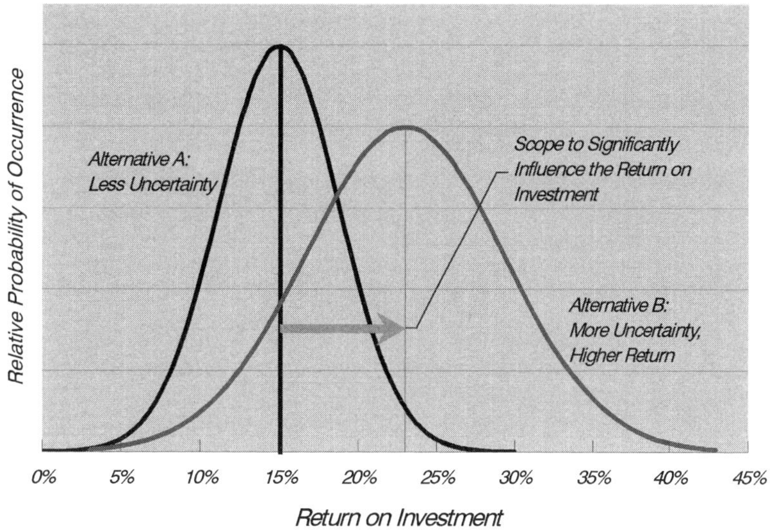

FIGURE 3.3 Return on investment during the strategic planning phase

when other mines cannot attract sales, or scope for adaptation as the mine is developed or as customer needs change. Most of these factors are intangible and do not show up on cash flow statements.

2. Development of the mine itself with sufficient scope for change. Most mines are developed around a selected base case, and the equipment sizing and development strategy are optimized for the base case estimate of market demand. Historically, this base case output usually proves to be too low in the long term (most mines expand from their initial production rate) but also too optimistic in the early stages of the mine life. With traditional investment decision methods, it is difficult to avoid adopting some base case production no matter how uncertain one may be of market interest. Planning and decision making have to be based on something! Nevertheless, the inevitability of change must be acknowledged. The strategy must plan for, and management must immediately start the implementation of processes leading to, changes in the mine in line with changes in the external world.
3. The adopted strategy that produces the preceding two results will yield a plan with a certain set of economic characteristics (cost structure, risk, etc.). The third objective of the strategic planning process is to understand which of these characteristics impacts the mine economics the most—as a guideline for following stages of the planning. Costs may then be optimized for those elements of the mine plan that are the most controllable.

Figure 3.3 shows in probabilistic terms the objective of mine planning during this strategic planning phase. Strategic planning is primarily concerned with changing, rejecting, and tentatively adopting choices that are fundamentally different in expected value terms. The focus is on choices that have the highest expected return, rather than on choices that are less risky. Although

Figure 3.3 presents these distributions as if they were *normal* distributions, in practice the dispersion of potential outcomes, even if it were known, is unlikely to fit any standard statistical shape. This does not invalidate the process—it merely recognizes the subjective nature of decision making with gross shortcomings in information. Thus, "expected value" in this context means the most likely outcome, or (in the case of a normal distribution) the mean of the distribution. Alternatives compared by using this focus of valuation are tentatively accepted on the following assumptions:

- Uncertainties (the spread of potential outcomes) can be reduced in subsequent evaluation phases.
- Subsequent evaluation will not significantly change the expected value of the distribution as uncertainties are resolved.

Once the strategic planning phase is complete, the overall expected return on investment seldom changes in any significant way. Subsequent phases of the planning are concerned with refining the plan.

If a detailed feasibility study has been completed and the project does not measure up to the required return on investment, then further detailed study is not the course of action to follow. A significant change in the return on investment requires some substantial or fundamental change in the strategy for development, not just refinement of some existing strategy.

Example 3.3:

At the start of many open pit gold projects, the mine is very shallow, grades are low, and mining costs as a proportion of overall project costs are very small. Most cost is in the ore treatment because of the large volumes of ore that must be handled. The priority for planning rests on ensuring delivery of the expected grade to the mill, even at the expense of mining costs. Any excess dilution will decrease throughput of ore proportionally and reduce revenue. The primary forms of technology a gold company needs at this stage are skills in ore treatment.

If, however, this first plan does not yield sufficient return, rather than fine-tuning the treatment process, the alternative may be to adopt an entirely different strategy. Development as an underground mine may allow much higher grades earlier in the mine life. In this alternative scenario, treatment costs may be the minor part of the mine cost structure. Mining costs will be more significant, and grade control will assume a lower priority than, say, reliability of ore delivery. Scheduling constraints, a common issue in underground mining, may be another priority. The primary forms of technology a gold company needs for this scenario are skills in underground mine development.

If a reexamination of the strategy is needed, it may be difficult to achieve with an in-house team whose skills have already been selected with an existing strategy in mind. The skills needed to understand alternative strategies (and implement them) may be completely different.

The strategic issues in a mine development can be broadly grouped in three ways, as shown in Table 3.2.

TABLE 3.2 Strategic issues in mine development

Strategic Factor	Example
Key strategic factors affecting revenue	▪ Selling price of product, including premiums or discounts for quality characteristics ▪ Quantity and quality of product produced ▪ Quantity of product able to be sold, including transport constraints and factors likely to disrupt mine output ▪ Start-up scheduling problems in the mine or mill delaying initial deliveries
Key strategic factors affecting cost	▪ Surface mines: Waste removal costs, rehabilitation costs ▪ Underground coal mines: Development to establish primary workings, roof support, ventilation costs ▪ Underground hard-rock mines: Development and access costs, drilling and blasting costs, backfilling costs, ground support ▪ Quarries: Fragmentation, environmental (blasting) constraints
Key strategic factors affecting risk	▪ Uncertainty in orebody definition (more critical in metalliferous mining, underground coal mines, and deeper mines) ▪ Price and currency volatility, as well as constraints associated with the financial structure limiting the scope for restructuring in the face of change ▪ Technology, as well as the potential for competitors to adopt new technology while the company's mine is constrained to use older, less economic technology ▪ Industrial disruption, as well as the constraints on change caused by institutionalized industrial agreements (when competitors are not so constrained) ▪ Environmental uncertainties (frequently the risk is not the *cost* of these imposts, but rather what additional costs are incurred because of delays) ▪ Political uncertainties: Access to land, community support

One of the key requirements of the strategic planning phase is to understand the risk characteristics of the particular mining method chosen. Subsequent planning can work out the most appropriate way of minimizing the impact of the uncertainty on the project cash flow.

Consider, for example, decision making for design of a processing facility. The economics of such a facility are sensitive to the precision of the estimates of plant yield, and often this cannot be determined accurately at the planning stage. Three strategically different alternatives present themselves:

1. Design the plant capacity for the "expected" plant yield. This alternative means that there is a 50% chance that the plant will be overdesigned. Capital will be spent that need not be spent. There is also a 50% chance that the plant will be underdesigned. In this event output will be reduced and revenue will suffer while all other costs remain unchanged. Is there scope for change in this plant design? Loss of revenue, since it is far more important than slightly higher cost, should be avoided even if higher costs are incurred. Overtime is an option available to some mine operators; it allows maintenance of production but at higher cost.
2. Design the plant capacity with production capacity to cover, say, 90% of the likely yield variation. This alternative uses capital to reduce risk. Compared to alternative 1, it is a decision to spend extra capital (a certain cost now) when on average there is only a 40% chance the extra capital will be needed. There is a 10% chance that the capital will still be inadequate.

From an engineering viewpoint, alternatives that have a 90% chance of success are preferable to alternatives that have only a 50% or less chance of success. From an economic viewpoint, *planning* to address the problem by using remedial measures may be cheaper.

3. Deliberately underdesign the plant (say, at a design capacity for which there is just a 30% probability that it will be adequate). At the same time, put in place contingency plans for immediate remedial action in the likely event (70% probability) that the plant indeed proves inadequate. The philosophy in this case is that the plant is expected to be inadequate, but it is perhaps not known in which particular area the inadequacy will show up, and the costs of covering all of the potential areas are excessive. Of course, there is always a 30% chance that it won't be inadequate at all.

Are there any guidelines as to which strategic option is preferable? The primary difference relates to institutionalized response to change.

An organization that has difficulty or high costs in accommodating change must favor alternative 2. The initial premium reduces the likelihood of subsequent change. On the balance of probabilities, this alternative results in overdesign—arguably still an "error" in economic terms—but frequently correctable by further expansion (including more capital if necessary).

An organization that can readily accommodate change has the potential to avoid expenditures that will automatically be made if alternative 2 is selected. Nevertheless, remedial measures are always very *evident* costs and are usually regarded as signs of failure, whereas the likely overcapitalization in alternative 2 is a less evident cost.

To summarize:

1. The primary economics of the mine are largely set in place by those factors that impact the revenue. The understanding of these factors (i.e., orebody characteristics) may be under the control of the mine planner, but the factors themselves may not be. The mine must be rich enough to be viable *even in the event of a low-probability outcome.* Alternatively, the plan must be adaptable enough to accommodate the changes necessary to be viable despite the uncertainty of the outcome.
2. Once the factors influencing revenues are addressed, the focus of the mine design must be on optimizing those factors that impact the cost structure of the mine the most. This is the area most under the control of the mine planner.
3. Mine planning must consider the risk factors—particularly those likely to cause loss of revenue. Mines can reduce their sensitivity to many risks through higher operating costs.

Many uncertainties cannot be removed by operating and capital cost changes to the plan or even by contingency planning. In these cases the decision to proceed is based on an "expected" return on investment high enough to cover the risk.

For those factors that *can* be quantified in capital or operating cost terms, a decision *not* to adopt the higher-cost/lower-risk alternative is a decision to accept a higher risk. In project assessments, the discount rate must reflect the risk. If it does not do so, then it implicitly biases the mine plan toward higher-risk options that yield a higher return.

An important outcome of this planning phase is to understand the cost structure of the mine as a guideline (the "rules") for future mine planning. The future mine plan must include an appropriate balance of

- overall focus on the more profitable reserves
- initial stages of the mine plan focused on reserves yielding strong cash flows (lower development costs, higher grades, etc.)
- mining schemes that facilitate change (expand, produce alternative products, etc.)

In open pit coal mining, the cost structure and related decisions are frequently determined through a cost-ranking analysis, described in more detail in Chapter 6.

In underground metal mining, a similar procedure usually involves cutoff-grade analysis and the influence of selling prices of polymetallic ores on development strategies. This is also covered in Chapter 6.

In open pit hard-rock mines, this phase of the planning process is called pit optimization. It defines the boundaries of a pit given certain conditions and the likely change in boundaries given changes in any of the inputs. This is also covered in Chapter 6.

The result of this strategic planning phase is to establish where mining is to take place, at what rate, and broadly how it is to be mined, i.e., what sequence and method. It establishes the ground rules for subsequent analysis.

Example 3.4:

Figure 3.4 shows two proportional pie charts of the cost structure of a large-scale open pit coal mine. The start of the mine, at a depth of 30 m, has relatively low costs—and 34% of these costs are in actual mining of the coal. At the end of the mine, at a depth of 100 m, the direct coal-mining costs represent only 9% of the much larger total. The dragline costs (representing 36% of the total costs) are also less significant than the costs associated with truck/shovel waste removal. At depth, for overall mine design efficiency, truck/shovel waste removal would take priority, even at the expense of dragline efficiency.

Subsequent Planning Phases

The contribution of the detailed and subsequent planning phases is to improve the *confidence* in the estimates—through understanding the implementability, constraints, and capital and operating costs. Figure 3.5 shows a probabilistic representation of the changes in return on investment during this detailed phase.

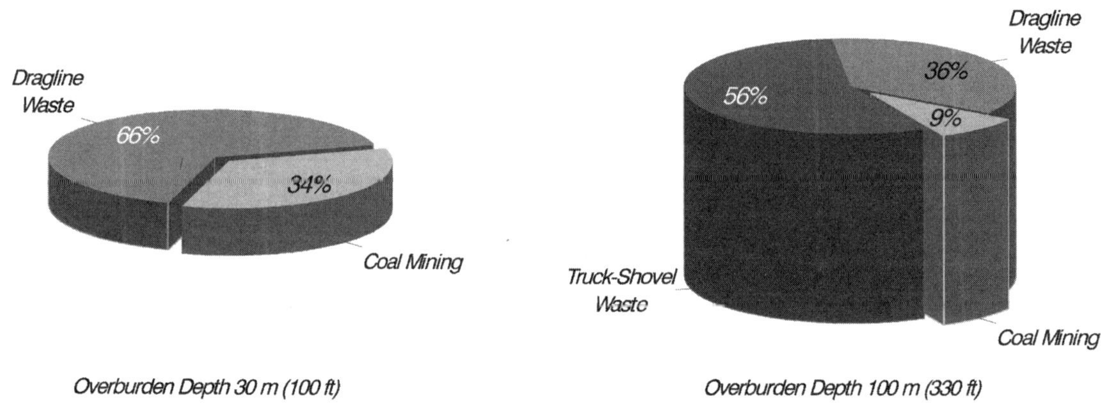

FIGURE 3.4 Cost structure through mine life

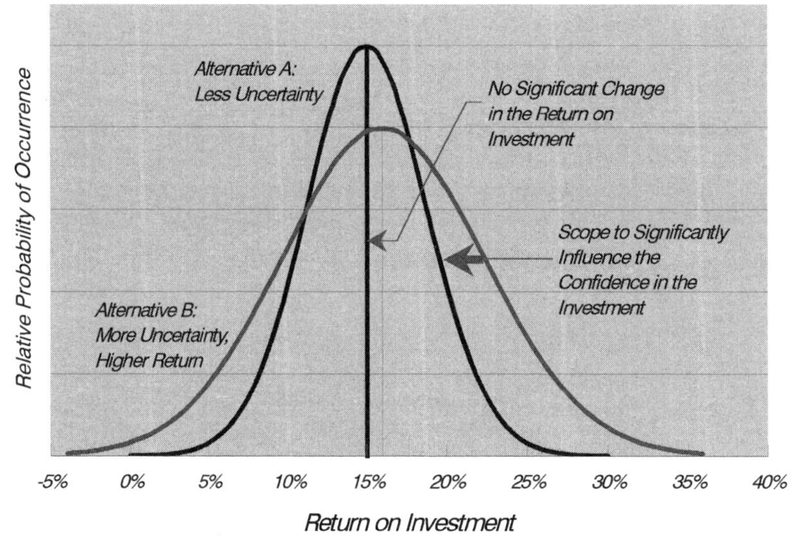

FIGURE 3.5 Effect on precision and return on investment during the detailed phase

Once the strategies are in place and the economics and risks associated with those strategies are clearly understood, then the details can be addressed in a long-term planning and feasibility phase. The key word for this and subsequent phases is *implementability*. The planning is directed at ensuring the highest probability that the production that is supposed to be achieved is in fact achieved. It also involves developing procedures so that this translation from "planned" to "actual" endures. The production techniques, costs, and cash flow characteristics will be greatly refined over those of the broad-brush phases, but it is unlikely that whole-mine economics (the whole-mine return on investment) will be notably different than the estimates in previous phases of study.

Issues to be addressed include

- equipment selection and optimization
- mine development, particularly the detailed layout of the mine and simulation of the revenue stream for the first few years, when the cash flows are most sensitive
- optimization of bench heights, widths, access constraints, and scheduling alternatives
- yearly, monthly, and shorter-term planning
- economic guidelines for short-term operational decision making

These issues are addressed in the balance of this book.

CHAPTER 4

Costs

To most people the concept of cost is subject to no ambiguity. It is the amount of money a person has to take out of his or her billfold, purse, or bank account to buy something. Day-to-day transactions seldom require any further discernment, since purchasing a small item on its own does not evidently preclude the purchase of anything else later on.

For larger items, the "real" cost is more evident. Purchasing a new car today might mean that a beach vacation next summer will be unaffordable—and the *real* cost of the car is the loss of enjoyment from the envisaged vacation that will no longer be possible.

In economics, the *cost* of anything is the highest-valued opportunity necessarily forsaken.

This chapter, indeed the whole book, looks at cost from this economic perspective. Unlike accounting costs, which are historical, the economic view of costs is a forward-looking one. (For a comprehensive treatment of costs and choice from an economic perspective, see Buchanan [1969; 1981] and Lewis [1948].) Costs in this sense inform decision making. If cost is to influence choice, it must be based on anticipations. *After* the fact, someone else might enjoy some of the benefits and endure some of the pain, but the choice is based on the anticipated value of this enjoyment or pain in the mind of the decision maker—*you*. The choice is between (1) this envisaged value (in *your* mind) if you choose one path and (2) the "cost" or the envisaged value if some alternative path is chosen.

This chapter illustrates the differences between common concepts of cost and the economic concept of cost. It introduces the concepts of marginal, average, variable, and fixed cost. It illustrates why the marginal cost calculation is such a vital one in pit optimization and in determining the scale of investments. It also illustrates why certain costs are excluded or included in cash flow calculations.

COST FROM AN ECONOMIC PERSPECTIVE

Failure to appreciate the purpose of the *economic* concept of cost may mean that efforts are misdirected. For example, the most common difficulty is confusion between the concept of "cost" and the undesirable attributes of some event. A new mine includes lots of undesirable attributes: the regulatory approval process, the wear and tear on local roads caused by the increased traffic, and the potential for construction difficulties. These are undesirable attributes of the mine, but they are not costs. (This section paraphrases a similar example from Alchian [1968].)

This temptation to think of the bad attributes as the cost is encouraged by business usage. In an evaluation of any proposition, revenues (the good consequences) are weighed against the expenses (the bad consequences). Thinking of the bad attributes as the cost overlooks the distinction between *valuation* and *costing*. The value of a given event or proposition is the sum of all of its elements, good and bad. A typical mining study assesses all of these good and bad attributes, and it determines a risk-adjusted, time-valued sum of revenues, operating expenses, taxes, and the like to arrive at a number commonly referred to as the net present value. The cost is the net present value of the next most attractive alternative proposition that is passed over in favor of the proposition at hand. The net present value of this alternative proposition is itself derived by weighing up the same good and bad attributes of that proposition.

Thus, it is incorrect to say that one of the costs of a new mine development is the deterioration of the local roads through increased traffic. Even if costs were looked at *only* in the state-of-the-roads dimension, the cost would not be the state of the roads after the mine goes ahead versus what they were before. Rather, it would be the *difference* between the likely state of the roads if the mine goes ahead and the likely state of the roads if the mine doesn't go ahead. If the mine doesn't go ahead, the reduced local taxes may mean deterioration of the roads in any case.

This economic concept of cost also has important implications for decisions regarding capital and other long-term commitments. Capital decisions are long-term decisions, but when the decision is made the choice is not necessarily an irrevocable path into the future. The decision to perform the action may be partly revocable, and it is only the irrevocable part that constitutes value or likely loss of value in the event of unanticipated obstacles to plan fulfillment. The cost of the *decision* (the irrevocable part) may not be the same as the cost of the *event*. It is the cost of the irrevocable part that is of concern. This concept is developed in a more comprehensive way in Chapter 15.

Example 4.1:

Assume that a dozer is to be used for reclamation and can be purchased for \$750,000. If the dozer lasts 4 years and can be sold for an expected \$75,000 after this time, is the cost \$750,000, \$675,000, or some other number?

Answer:

If the decision included an irrevocable commitment to retain ownership of the dozer until the end of the 4-year period, the cost would be $675,000. (Actually, the receipt of $75,000 in 4 years time, assuming a 10% rate for the time value of money, would be $75,000/(1.10)^4 = $51,226; thus, the monetary cost in present value terms is $698,774.) But presumably the decision is not irrevocable, in which case the dozer can always be sold again–maybe next week! If the dozer could be sold next week for an expected $700,000, then the cost of the *decision* is only $50,000.

Example 4.2:

Assume now that a contractor can undertake the same dozer work for a payment of $2.00 per unit of production, achieving the same rate that would be achieved if a dozer were purchased: 10,000 units of production per week. What is the cost of this option?

Answer:

Because this style of contract work is remunerated on a *per unit* basis, there is a tendency to consider costs only as a direct function of production. But the same logic as set out previously applies here as well. It is the term of the commitment that matters, as well as the scope at the end of this term to continue or abandon the task. If the contract were terminated next week, after only 10,000 units had been moved, would the payment to the contractor amount to only $20,000? Most contracts have additional payments in the event of early termination, and the real cost of this option may be substantially higher. The cost of the decision to use a contractor may be more than the cost of the decision to purchase the dozer.

These simple illustrations also demonstrate an important difference between the accounting treatment of costs and the economic treatment of costs. The accounting treatment is based on historical measures independent of options or follow-on choices during the life of equipment. Thus, depreciation from an accounting perspective typically means a constant reduction in value of the dozer over time. From an economic perspective, depreciation within any period of commitment means the difference between (1) the value of the dozer if that value had to be realized at the start of the period and (2) the value if it had to be realized at the end of the period. This "economic depreciation" informs much decision making in the mining industry, from investments in access developments to investments in new technology when the older equipment is not worn out.

In the balance of this chapter, the strict distinction between the "cost" of something (the value of some alternative foregone) and "cost" in common usage (the monetary expense) will henceforth be overlooked. Except for large investments, the monetary expense *is* quite a good proxy for the value of the foregone alternative in a market economy. Yet the distinction should not be forgotten. Mining companies working in less developed countries, for instance, cannot assume that market prices are an appropriate proxy for the value of foregone alternatives. If a special O-ring for an important pump is unavailable, then the cost of the O-ring is definitely *not* the $5 catalog list price—it may be thousands of dollars of forgone production.

The distinction between the market value of something and the (private) value of the same thing is also important in capital investment choice. The market value of a mining company in the midst of a major mine development program may not be faithfully represented by the company's share price. Faithful market valuation requires an informed market, and in the process of developing mines there are times when markets are necessarily quite ill informed.

TYPES OF COSTS

Every business needs to know what the costs to produce its products are if it is to make sensible business decisions. There are a variety of ways to present and apply costs, and some cost concepts are more appropriate for certain problems than others. This section introduces some of the more important cost concepts, including

- fixed costs
- sunk costs
- recoverable costs
- opportunity costs
- variable costs
- operating costs
- externalities

and explores some of the subtleties in understanding them. (This section draws primarily from Carlton and Perloff [1994, p. 51].)

Every business incurs costs that do not vary with output, as well as costs that do. A fixed cost is an expense that does not vary with the level of output. Annual payments to maintain a mining lease (assuming the payments are independent of production) is one example of a fixed cost. The construction cost of a high-voltage power line into a mine site is another fixed cost.

The portion of a fixed cost that is not recoverable is a sunk cost. Sunk costs should not affect subsequent decisions. In the preparation of a cash flow of a mining property, sunk costs are excluded.

Example 4.3:

You have spent $15 million evaluating a mining property over a long period of time, and the project looks (almost) viable. Your accounting policy requires you to allocate the $15 million across the proven reserves, and when this cost is included the project fails to meet your required investment return. Should the exploration costs be included or excluded?

Answer:

The exploration costs should not enter into the decision to proceed or not. If you proceed with the project, your accountants will be reporting a "loss" on the project (because they will be writing off the high cost of exploration and assigning it to the project), but if you don't proceed your accountants will still be reporting a "loss." The $15 million is common to all alternatives because it has already been expended. However, some exploration cost may be recoverable (as discussed later).

The exploration costs in the preceding example may have already been spent, but they do not automatically become sunk costs. The whole $15 million of expenditure might not be able to be recovered, for example, but the property might be saleable for $10 million. In this case, only $5 million of the original $15 million is a sunk cost, and $10 million is a recoverable cost.

How does a company treat the costs of things that it already owns? The property in Example 4.3 (which cost $15 million to explore and that could potentially be sold for $10 million) is another example. The associated "costs" are true economic costs as described in the preceding section of this chapter because they are defined by the value of the opportunity that is forsaken. In the finance literature, these true economic costs are commonly referred to as opportunity costs.

The key to understanding opportunity cost is not "before versus after," but rather "with versus without" (Brealey and Myers 2003, p. 121). If a company already owns a machine and applies it to some new task, then the ownership (as well as the cash flow) associated with the machine is the same both before and after use of the machine in the new task. However, if the company undertook the new task *without* using the machine, how would the cash flow compare to that when the new task is undertaken *with* the machine?

Example 4.4:

You have some old equipment that cannot be used for overburden removal, and you propose to use it for reclamation. You already own it, so there is no purchase price and no cash flow. If you do not use it for reclamation, you could sell it for $1 million. Should the $1 million be included in the cash flow analysis and in the decision to use the equipment for reclamation?

Answer:

Yes! The "with" case—call it case A—involves undertaking the reclamation with the already-owned equipment. The "without" case—call it case B—involves undertaking the reclamation by some other means without the equipment and selling the equipment. Case B has its own costs, plus a revenue of $1 million (minus taxes) from the sale of the equipment; if case A proceeds this potential revenue is lost.

Lost revenues (from the alternative scenarios) are called opportunity costs because, by accepting the project, a company foregoes other opportunities for using the assets.

Variable costs are costs that change with the level of output. Typically, as output increases, so does the need for labor, fuel, electricity, and materials, so variable costs depend on the wages and prices that a firm must pay for these inputs.

Although these variable costs are commonly called operating costs, the decisions that are made day-to-day in a mine cannot assume a one-to-one correspondence between what an accountant calls an operating cost and what is truly a variable cost. Whether a cost is fixed or variable depends upon the time frame of the decision. For yearly budgeting, labor costs are a variable cost because labor requirements can be increased or decreased in line with yearly production requirements. However, for day-to-day decisions by a mine supervisor, even labor costs might be fixed. If a truck driver has reported to work and there is no truck available, then this labor cost cannot be avoided. Chapter 10 sets out a case study analyzing fixed and variable costs and avoidable costs as a function of the time frame of the decision maker.

Any new mining development also includes costs that the decision maker does *not* take into account. Following commencement of a new mine, the increased traffic might require higher costs of local road maintenance, for instance. Dust and noise pollution might impose costs on people quite removed from the project. This type of cost is termed an externality. Externalities can be both positive and negative. A supermarket valued at $0.5 million before a mine commences might be valued at $1 million after the mine starts because of the increased patronage it enjoys from mine personnel. Externalities are changes in value that are borne by others and are not taken into account in the decision.

Efficiency and the correctness of choices in some society-wide sense suggest that decisions should take all of these externalities into account. Many large firms already do this on "social responsibility" grounds even if there is no legislated requirement to do so. Local taxes compensate for increased road usage, leaving existing residents no worse off than before. In some jurisdictions, markets exist to purchase the rights for gaseous emissions into the atmosphere and for sediment runoff into water catchments. These mechanisms allow previously uncounted costs to be internalized. At the same time, mining companies can sometimes enlist the assistance of (i.e., internalize some of the advantages from) local enterprises, such as supermarkets and transport companies, who stand to gain from new development.

MARGINAL COSTS

In economics, few concepts are more important than the concept of marginal cost.

The marginal cost is the change in total cost. The counterpart to marginal cost is marginal revenue; i.e., marginal revenue is the change in total revenue.

Consider almost any production process. The process will involve some fixed costs and some variable costs. As production expands, the fixed costs are unchanged, so the average per-unit cost of production attributable to this

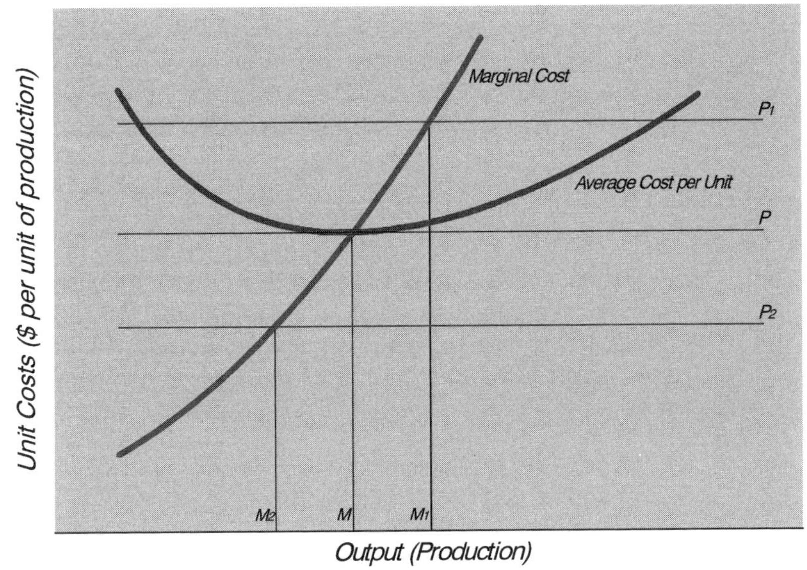

FIGURE 4.1 Average and marginal cost curves

component declines. If this was the only trend, then the highest-production case would represent the lowest overall average cost of production.

Few production processes work this way, however. The "fixed" parts of the process can service only a limited range of variable parts. As production expands for the same fixed components, the efficiency of the system declines. Each increment of production incurs variable costs a little more than the previous increment.

A truck/shovel system is the archetypal system in mining. The shovel is the fixed component, and the trucks are the variable component. When only *one* truck is paired to the shovel, the average cost of production is high because the fixed costs of owning and operating the shovel are spread over a relatively small production. When two trucks are allocated, production will increase–but not quite to double the previous amount because there will be queuing at the start of the shift. As additional trucks are added, production will increase but by a declining amount as the increasing numbers of trucks interfere with each other. (A comprehensive case study using this truck/shovel example is set out in Chapter 10.)

Another example is the railway connecting a mine to a port or to the market. The fixed costs of the track can service a lot of trains, but each additional train adds scheduling complexity and increased delays into the whole system. With expanded production, more and more work has to be done at higher-cost overtime rates. With continued expansion, it eventually become more efficient to duplicate the track.

Figure 4.1 shows the idealized situation for this style of production process. In the figure, the costs shown include allowance for repayment of capital. The average cost of production is high at low levels of production, and each increment of production has a low but increasing marginal cost. If the marginal cost is less than the average cost, the average cost declines with increases in production. The output rate (M) that yields the lowest average unit cost of production occurs where the marginal cost curve crosses the average cost curve.

The lowest average unit cost of production is certainly a desirable objective, but usually the objective is to maximize profits (or minimize losses). If the selling price is P_1, for example, production can be expanded to M_1, and the additional production still yields a return higher than the marginal cost. Indeed, this is the rule: Expand production until the marginal cost equates to the (marginal) return.

In Figure 4.1, if the selling price is P_2, losses cannot be avoided at any output level. However, the losses are minimized at the output level M_2 where the marginal cost equates to the price P_2, not at point M.

In this example, the price (P, P_1, and P_2) was assumed to be independent of production. For many mineral commodities, such as gold and silver, this is an appropriate assumption. For many other commodities, though, price is also dependent on production, and additional production can be absorbed in the market only if prices reduce. Most industrial minerals are in this category, as are bulk commodities like coal. How, then, can the optimum output be determined?

In these cases (where price cannot be assumed constant), the same logic applies–select an output level where the marginal cost equates to the return. In this case the return is not the "constant" price but rather the marginal revenue. Example 4.5 illustrates how the marginal return calculation can be addressed.

Example 4.5:

Consider a mine currently producing 8,000,000 t of coal per year under a mix of spot sales and contracts to a variety of regional customers. At any one time, customers are all paying slightly different prices for the same coal, but in due course prices become more widely known and these influences reduce. Even the long-term contracts have price variation clauses that adjust to market conditions. The average selling price for the existing output is \$10/t. You can expand production by 1,000,000 tpy at an operating cost (for this extra coal) of \$6/t with only a small amount of capital. Repaying this incremental capital, you can still make your required return on investment at a selling price of only \$9/t. You believe you can find additional customers who will purchase the extra 1,000,000 tpy at \$9/t. Should you proceed?

Answer:

This seems to be a clear-cut case. If the price exceeds the marginal cost, profits increase with each increase in production. The risk is that, lacking any ability to keep prices secret (never a very good strategy, in any case) for any length of time or to differentiate the "new" coal from the "old" coal, all of the output

TABLE 4.1 Marginal revenue calculation for Example 4.5

Production Scenario	Production (tpy)	Estimated Selling Price ($/t)	Annual Revenue ($)
Current mine	8,000,000	10.00	80,000,000
After expansion	9,000,000	9.50	85,500,000
Extra production and revenue	1,000,000	5.50* *unit marginal revenue*	5,500,000 *marginal revenue*

* Unit marginal revenue ($5.50/t) determined from (marginal revenue)/(marginal production)

> from the mine will be priced downward. Your customers themselves have incentives to do this. For instance, what is to stop your new customer from selling some of your new coal to one of your old customers (at perhaps $9.50/t), with both of them being better off?
>
> In this example, prices are not independent of production. Selling some coal below average price makes it harder to maintain the price of your existing supply. Perhaps you risk downward price revision by only 5%—but this reduced price applies to all of your output, if not immediately, then certainly in the near future. The marginal revenue is not $9/t but rather the change in total revenue. Expansion is viable only to the point where marginal cost equates to marginal revenue. Table 4.1 shows this calculation. The additional output has an effective selling price (marginal revenue) of just $5.50/t—a price at which the expanded production is not viable.

The confusion between average and marginal payoffs can work the other way, too. Most managers naturally hesitate to throw good money after bad, but if an existing project is already making a loss, existing losses may be irrelevant in the decision regarding incremental expenditure on the project. Sometimes an existing project is yielding poor returns because of a bottleneck in the production chain. Small incremental investments to remove such bottlenecks can yield large marginal returns.

Whenever optimization is the objective, marginal costs should be the focus. Procedures aimed at pit optimization apply this identical principle. Starting from some initial orebody, they examine extensions to the orebody in all dimensions to ascertain whether the marginal revenues from the extension exceed the marginal cost of extracting the additional ore and waste. The optimum pit is the one where, at the margins, the return equates to the cost. Chapter 6 sets out examples of applying this principle for several different types of orebodies.

COSTS WITH MULTIPRODUCT MINES

Many mines produce more than one product. Most copper mines produce by-product gold and silver. Silver, lead, and zinc commonly occur together, though in orebodies that may contain varying proportions of each. How can the concepts of economics be applied faithfully to such multiproduct enterprises?

If a mining enterprise produces two or more products, there is no one average cost or even one marginal cost because there is no one measure of output. (Some economic tools have been devised to handle parts of this problem; however, in the opinion of the author, there is limited scope for application of these tools in a practical mining environment. See, for example, Baumol, Panzar, and Willig [1982] and Shepherd [1984].) There are, however, some cost concepts that are analogous to those in a single-product environment.

Most multiproduct mines use a proxy for mine output by weighting the products by their respective selling prices. Thus, silver-lead mines might use a lead-equivalent output by adding to the lead output the total tons of lead that would yield the same revenue as that received from the silver. In an era before computers were widespread, this was a useful operational tool for mine planners who had long experience in mine economics based on the grade of a single type of ore. In light of increased computational capability, it is uncertain if such a construct aids analysis significantly enough to be worth the effort. With computers it is simpler to assign total revenue (and total costs) to mining blocks rather than some pseudo ore grade. In any case the total revenue must first be calculated in order to determine the single ore-equivalent (pseudo) grade.

In multiproduct mines, the marginal cost of a single product is somewhat easier to define. For example, if Q_1 tons of lead and Q_2 tons of zinc are produced, the marginal cost of producing lead is the additional cost incurred in increasing Q_1 to $Q_1 + 1$ while holding the output of zinc constant at Q_2. Nevertheless, such a concept is difficult to apply in practice. In practice the decision that must be made is whether to expand or not, and this decision rests on the marginal revenue from additional production exceeding the marginal cost. An orebody that contains only lead can be used to readily deduce this marginal cost of production for lead, but this may not be the cost that should be used for decisions. The most profitable way of producing additional lead may be to expand production of both metals. In this case, the marginal cost is the change in total cost, and the marginal revenue includes the revenue from selling the additional zinc.

CHAPTER 5

Time Value of Money

Money tomorrow is not as valuable as money today. Given the choice of having the same amount of money in the future or right now, everyone would prefer to have it now.* Money to be received in the future might not materialize. Even if there was no such risk, money is still worth more if it is available for use now. If it is available *now,* the opportunity set of possible uses for the money is as broad as possible. If the money won't be available until some time in the future, then the opportunity set is limited to a smaller subset of this first set. Maximum freedom to choose is always worth something–particularly in more uncertain environments.

Therefore, money to be received in the future must include a premium if it is to be thought of as being equivalent to money today. Future cash flows (i.e., money) must be discounted to be compared with cash flows in the present.

Almost every economic decision in mining involves cash flows (spending money, receiving money) occurring at different points of time. Consequently, economic evaluations must incorporate a way for equating these money values at some constant point in time (usually, now).

For simple calculations, future values (FVs) are calculated by taking current values and multiplying by the compound interest rate. Or, equivalently, future values (anticipated cash flows) are turned into the equivalent present value (PV) by discounting, i.e., by being divided by the compound interest rate.

For meaningful calculations of mining investment propositions, a complete tabulation is normally prepared for all of the cash flows occurring through each year of the life of the project. The aggregate cash flow (the sum of the expected positive and negative cash flows) occurring in each year is calculated first, and then this value is turned into a present value via the applied discount

* The same is true of most commodities, but other commodities usually have higher holding costs and require greater transaction costs to turn them from their present state into the state in which they are most useful. Thus, perishable items can frequently have a higher value for future delivery than for immediate delivery.

rate. The discount rate is usually greater than the long-term interest rate to account for uncertainty and other factors.

The first main section of this chapter sets out the general process of discounting, determining present value and future values, and determining constant annual amounts that equate to a certain present value. The next main section sets out the application of the time value concept and use of discounted cash flow methods. The remaining sections expand on the discussion to include inflation effects, DCF ranking criteria, and the concept of a discounted average cost.

VALUATION AT A CONSTANT POINT IN TIME

The first series of time-value calculations apply simple formulas to bring anticipated cash flows to an equivalent time reference basis for calculation. Simple calculations are grouped into two categories, namely

- how to turn a future value into the equivalent present value and vice versa
- how to turn a regular series of equal values occurring over several years into an equivalent single amount in the present and vice versa

Present Values and Future Values

The two functions used to relate present values to future values and vice versa are

- the compound amount function (future value)
- the present value function

The future value (compound amount function) is determined by the following formula:

$$FV = PV \times (1 + i)^n \quad \textbf{(EQ 5.1)}$$

where

FV = the future value

PV = the present value

i = the interest rate (in the time period)

n = the number of time periods (years)

$(1 + i)^n$ = the compound factor

Example 5.1:

Your company has to pay a reclamation bond to the government for each hectare of land disturbed. The funds are held in trust, earning interest at 6% compounded annually, until reclamation is complete, whereupon they are returned. If you disturb 40 hectares of land this year and the bond is \$50,000 per hectare, how much do you expect to get back when reclamation is completed in 3 years' time?

Answer:

Present value (of money paid out now)	\$50,000 × 40 = \$2,000,000
Compound factor	$(1 + 0.06)^3 = 1.191$
Future value	\$2,382,000

The present value function is used to move a future value estimate back to the present; it is the inverse of the future value function:

$$\text{PV} = \text{FV}\left[\frac{1}{(1+i)^n}\right] \quad \textbf{(EQ 5.2)}$$

where

PV = the present value

FV = the future value

i = the interest rate (in the time period)

n = the number of time periods (years)

In Equation 5.2, the term $[1/(1 + i)^n]$ is known as the present value function.

Example 5.2:

You have received bids from two manufacturers for purchase of a new dragline. The first bid is very competitive but is from a company that requires payment in full on placement of an order. The second bid is for a higher price, but no payment is required until the machine starts digging 3 years hence. Which is the preferred option? If the first dragline is purchased, what is the effective return on investment for the 3 years?

Dragline A bid price (payment today)	\$30,000,000
Dragline B bid price (payment in 3 years' time)	\$40,000,000
Required return on capital (discount rate i)	15%
Taxation	Ignore for this example
Time (n)	3 years

Answer:

Present value of purchasing dragline B 3 years into the future	\$40,000,000 × 0.6575 = \$26,300,000

Thus, dragline B has a lower cost than dragline A in present value terms.

Compound factor for investment in dragline A versus dragline B	\$40,000,000/\$30,000,000 = 1.3333
Effective return on investment over 3 years	$(1 + i)^3 = 1.3333$

$$i = 1.3333^{0.3333} - 1$$
$$= 10.06\%$$

Almost all scientific or engineering calculators are now equipped with functions capable of undertaking the preceding calculations directly; this is the recommended way of determining the compound amount and present value functions. A table of present value factors and compound amount factors is also included in Appendix A.

Regular Streams of Income and/or Expenses

The two functions used to relate a regular series of equal values occurring over several years into an equivalent single amount are

- the capital recovery function
- the sinking fund function

These functions are a little more complicated to calculate than the present value and future value functions, since they involve a series of payments over several time periods.

Capital Recovery Function The capital recovery function is used to spread a present value amount out evenly over a period of n years. It produces a series of equal values occurring at the end of each year for the time period specified.

Consider first the simple case. Paying off a loan after 1 year in one installment requires a payment of

$$\text{payment} = \text{loan amount} \times (1 + i)^1$$

Paying off a loan in 2 years by two equal installments at the end of each year requires an annual payment of X, where X is determined by the following expression:

$$[\{[\text{loan amount} \times (1 + i)] - X\} \times (1 + i)] - X = 0$$

In the general case:

$$\text{capital recovery factor} = \frac{i}{1 - \frac{1}{(1+i)^n}} = \frac{i(1+i)^n}{(1+i)^n - 1} \quad \textbf{(EQ 5.3)}$$

The capital recovery factor is multiplied by the initial total cost to yield the equivalent annual cost. Example 5.3 illustrates how the capital recovery factor is used in a calculation. A table of capital recovery factors is included in Appendix A.

Example 5.3:

The expected life of a rope shovel is 16 years, after which time the mine will close and the salvage value will be effectively zero. What is the annual owning cost, including allowance for return on your capital invested in the rope shovel? If the shovel works 6,000 hours per year, what is the hourly cost?

Required return on capital (discount rate *i*)	15%
Taxation	Ignore for this example
Cost of rope shovel	$7,000,000
Time (*n*)	16 years

Answer:

Capital recovery factor (from Equation 5.3 or Appendix A)	$0.15/[1 - (1/1.15)^{16}]$ $= 0.15/[1 - 0.1069]$ $= 0.1679$
Equivalent annual "cost" of shovel over 16 years	$7,000,000 × 0.1679 = $1,175,300/year
Hourly cost	$1,175,300/6,000 = $195.88/hour

Sinking Fund Function Just as the capital recovery factor equates a present value with a series of equal installments over time, the sinking fund function spreads a future value evenly over time. It is used to provide for a known expenditure at some time in the future by setting aside equal annual amounts each year starting from the present. The most common use of this sort of calculation is for retirement planning. A typical question might be: If someone wants to retire in 25 years with a $500,000 lump sum, how much money needs to be placed in a fund per year to achieve this amount at the end of the 25 years?

The sinking fund function is calculated using the following formula:

$$\text{sinking fund factor} = \frac{i}{(1+i)^n - 1} \qquad \textbf{(EQ 5.4)}$$

The future value is multiplied by the sinking fund factor to determine the equivalent annual payments. Example 5.4 illustrates how the sinking fund factor is used in a typical calculation. A table of sinking fund factors is included in Appendix A.

Example 5.4:

A contractor has purchased a new hydraulic excavator for a 5-year-life job. He expects the excavator to last 8 years but knows that he will also have to perform a major overhaul on it costing $1,000,000 at the end of the 5-year period. The overhaul cost will be built into the bid price for the job, with annual payments being placed in a sinking fund invested at 8% compounded per year. If the excavator works 3,000 hours per year, how much should be budgeted (and set aside) per hour to provide for the major overhaul?

Sinking fund return on funds	8%
Taxation	Ignore for this example
Cost of major overhaul	$1,000,000
Time (n)	5 years

Answer:

Sinking fund factor (from Equation 5.4 or Appendix A)	$0.08/(1.08^5 - 1)$ $= 0.08/0.4693$ $= 0.1705$
Equivalent annual amount to set aside	$1,000,000 × 0.1705 = $170,456/year
Hourly cost	$170,456/3,000 = $56.82/hour

The four functions given by Equations 5.1 through 5.4 are used in comparing alternatives and determining values for activities occurring over several units of time (e.g., years) whenever the taxation effects are likely to be minimal.

DISCOUNTED CASH FLOW ANALYSIS

All of the functions discussed in the preceding section are important in determining values for activities occurring over time. However, their usefulness is limited by the fact that they do not take taxation effects into account and they need regular cash flows. Since almost all real-life cases involve taxation and since operating costs and revenues vary over time, an alternative evaluation method must be used. The method universally used for almost all mining and other business evaluations is the discounted cash flow technique.

Basis of Cash Flow Analysis

There is a big difference between corporate finance (i.e., costing, economics, and capital investment decisions), which is addressed in this book, and financial accounting, which stresses incomes and earnings. Accounting procedures document what has happened. Mining economics aims at informed decisions on *what to do.*

For *accounting* purposes, all expenditures are normally apportioned over the period of useful work. For planning and operating a business, there is no apportionment—allowance has to be made for real cash inflows and outflows when they actually occur. Example 5.5 illustrates the difference.

Example 5.5:

Consider the purchase of a dozer for $600,000 paid for today. The entire $600,000 is an immediate cash outflow. An amount of $600,000 has to be available from somewhere at the time the dozer is delivered—before it has done any useful work. However, assuming straight line depreciation over the 6-year life of the dozer, only $100,000 is considered an accounting expense in the current year. Current earnings (reported profits) are reduced by only $100,000. The remaining $500,000 is expensed (counted as an operating cost) over the following 5 years.

To run a business, what is important is cash flow, not "accounting" profit. In Example 5.5, the company supplying the dozer requires the full purchase price to be paid now, not just the amount of depreciation that the accountant attributes to this year's cost of production. Furthermore, capital expenditures always occur *before* any production, whereas accounting conventions assign their costs (and revenues) only during or after production has taken place.

Cash flow analysis involves simulating what is happening or what is expected to happen in the mine over time. It is a forward-looking, or *ex ante,* process. All cash flows—money flowing into or out of the company bank accounts—are included.

The purpose of preparing a cash flow is to be able to make a decision. The extra cash flow associated with the company's investing in a particular project—call it project A—is compared to the cash flow associated with whatever else might be chosen. Call this alternative investment project B. Though this fact is not always evident, there is *always* a project B. Project B may simply entail leaving the money in the bank, or declaring a higher dividend, or buying back one's own shares. A proposed 20-year-life mine is economically attractive or unattractive only in relation to what else might be done in the ensuing 20 years. There is no stopping of time—the decision is only a comparative one concerning alternative paths into the future. If a cash flow is common to both options, then it has no real effect on the decision—a company is interested only in the changes in the company cash flow that occur as a direct consequence of accepting the project.

In the preparation of cash flow tabulations, there are important conventions that *do* follow accounting principles. One of the most important of these is the end-of-year convention. This convention (which isn't inviolate but is certainly the default rule) tabulates all cash flows as occurring at the *end* of the year in which they actually occur. Revenues from the sale of the mine output are all assumed to occur at the end of the year, as are the mine operating expenses. A machine that is working for an entire year would have all of its operating costs assumed to occur at the end of the year. However, to be working through the year, the machine itself would have to be paid for and operational at the very start of the year. Therefore, under this convention, capital expenditures occur at the start of the year and are assigned to the (end of the) previous year. This convention is also the one followed in the capital recovery function described earlier in this chapter.

There are other conventions, and some companies follow a middle-of-year convention for certain activities. For most mining project evaluations, the precision of the end-of-year convention is entirely adequate as a basis for decision making and long-term cash flow requirements within the company.

There are a number of other accounting conventions, principally relating to the calculation of tax, as well as the tax treatment of certain types of assets. These conventions will be easier to describe after examination of a typical cash flow analysis.

A Sample Discounted Cash Flow

The primary technical tool used by firms worldwide for assessment of capital investments is the discounted cash flow technique, the traditional application of which is described next. Assuming the output is being sold into an existing competitive market, the steps involved in setting up the cash flow and for using the results are as follows:

1. The cash flow tabulates what is happening over time, and the columns on the top of the spreadsheet represent years (usually) extending from the present for as long as the analysis is relevant.
2. An assessment is made of the likely selling prices of the product(s) in the general marketplace for the period under study.
3. Production quantities and capital and operating costs for the life of the project are estimated by technical personnel.
4. The output from the mine multiplied by the selling price is the source of revenue–normally entered on the first rows of the spreadsheet. Most mines produce only a small number of products, so the quantity is normally tabulated on an annual basis for each different product or group of products. There may be other revenues, although these are typically small. Other revenues include the resale or salvage value of equipment, as well as the return of reclamation bonds and the like.
5. There are usually three different types of cash outflow: capital expenditures, operating expenses, and taxes. Operating expenses are usually built up separately by technical personnel and enter the spreadsheet along with operating revenue to determine operating profit.
6. Capital expenditures are clear cash outflows entered in the spreadsheet in the year in which they occur.
7. The tax calculation is a major component of the cash flow. It is described in more detail later in this chapter (see "Depreciation, Depletion, Tax Credits, and Taxation"). Components included in the cash flow include equipment depreciation for tax purposes, depletion allowances, calculation of taxable profit for both state and federal purposes, and tax. Some taxes may or may not be allowable deductions in determining the taxable income for other taxes.*
8. The net cash flow is the yearly sum of all of the cash inflow minus all of the cash outflow. This is the actual net amount of money expected to flow into or out of the company bank account in any given year.

* For example, state taxes may be allowable business expenses in the assessment of liability for federal taxation. Alternatively, taxes for so-called fringe benefits, paid by employers for "benefits" offered to employees, may not be an allowable expense in the assessment of liability for federal taxation. Since taxation laws vary widely from country to country and over time, in all such cases a competent taxation authority should be consulted before estimated taxation liabilities are finalized.

9. The firm assesses, or is aware of, the return that its capital and other resources are likely to yield if placed in alternative projects, returned to shareholders, or kept in some more liquid form. This establishes the opportunity cost of capital, a return that has to be met or exceeded for any new project to proceed. Cash flows are discounted at this rate according to Equation 5.2 to determine the net present value of the project.

Consider, for example, a hypothetical gold mine with a 5-year life. The main background data needed for a discounted cash flow tabulation of this simplified mining project are set out in Table 5.1. Such background data are needed for any discounted cash flow tabulation. Table 5.2 shows the idealized DCF tabulation.

In Table 5.2, the data have been deliberately chosen so that the selling price of \$500/oz yields a net present value of zero at a discount rate of 15%. The internal rate of return (IRR)—the discount rate for which the net present value is zero—is therefore 15%. Particular characteristics of Table 5.2 and their similarity to generalized cash flow calculations are described in the following sections of this chapter.

Capital Expenditures, Production, Revenues, and Operating Costs

The objective of a cash flow analysis is to simulate all of the anticipated cash flows for the project over its life (and express them in present value terms) so that a decision can be made. The most obvious cash flows are

- revenues from sale of the products
- expenses incurred in producing the products
- capital expenditures necessary to bring about the production

Capital expenditures are tabulated in the cash flow in the year *prior* to their use. The plant or equipment must be operational before any production takes place (the start of the period). Therefore, under the end-of-year convention, capital expenditures are placed at the end of the preceding year.

Cash flow tabulations should normally commence with production tabulated at the top or near the top of the table, since almost all of the revenue and

TABLE 5.1 Base data for discounted cash flow calculation

Item	Value
Initial capital cost	\$15,000,000
Life of project	5 years
Salvage value at end of life	At written-down value (see example in Table 5.2)
Production per year	Varies as shown in Table 5.2
Selling price	\$500/oz
Annual operating expenses	As shown in Table 5.2
Depreciation rate for tax purposes	27.5% (declining balance method)
Tax rate	35%
Discount rate	15%

TABLE 5.2 Sample discounted cash flow

	Year					
	0	**1**	**2**	**3**	**4**	**5**
Production, *oz*		30,000	50,000	50,000	50,000	45,000
Operating revenue at **$500/oz**, *thousand $*		15,000	25,000	25,000	25,000	22,500
Operating expenses, *thousand $*		10,598	17,762	19,339	21,073	20,882
Operating profit, *thousand $*		4,402	7,238	5,661	3,927	1,618
Capital expenditure, *thousand $*	15,000					
Tax depreciation—27.5% of the start-of-year value of capital, *thousand $*		4,125	2,991	2,168	1,572	1,140
End-of-year written-down value for tax purposes, *thousand $*		10,875	7,884	5,716	4,144	3,005
Salvage value, *thousand $*						3,005
Taxable profit, *thousand $*		277	4,247	3,493	2,355	478
Income tax payable at 35% tax rate, *thousand $*		97	1,486	1,223	824	167
After-tax profit, *thousand $*		180	2,760	2,270	1,531	311
Net cash flow, *thousand $*	(15,000)	4,305	5,751	4,439	3,103	4,455
Discount factor (at 15% return on investment)	*1.0000*	*0.8696*	*0.7561*	*0.6575*	*0.5718*	*0.4972*
Discounted cash flow, *thousand $*	(15,000)	3,744	4,349	2,919	1,774	2,215
Net present value, $	**0**					

Note: Numbers in parentheses—for example, "(15,000)"—represent negative values.

many of the operating expenses are related to production. Revenues (the primary cash inflow) are also tabulated at the top of the table.

Operating costs are subtracted from the operating revenues to obtain the operating profit. *All* operating expenses are included in a cash flow calculation, even if some of the costs pertain to following years' production. This differs from the way operating costs are treated for accounting purposes, for which expenses that pertain to production in succeeding time periods (e.g., advance stripping) are apportioned to the period in which they directly relate to production.

Working Capital, Salvage Values, and the Timing of Cash Flows

Most capital expenditure relates to fixed assets that contribute directly to production–equipment, buildings, and the like. The initial purchase of these capital items is a cash outflow, usually occurring at the start of the project. Operating expenses such as repair parts are also a cash outflow and must be paid for at the time they are used.

In practice, there are always timing differences between when costs are incurred (or revenues received) and when the spare part is used or production undertaken.

Example 5.6:

When a repair part is needed for an essential item of mining equipment, the maintenance staff take one out of the store. Only after that does the warehouse order the part, wait for it to be delivered, and pay for it. For practical purposes, the operating expense is incurred as the part is used, except for the initial cost of purchasing the initial inventory of spare parts for the warehouse.

Working Capital The initial inventory of spare parts is one example of working capital. In a detailed cash flow analysis, the initial stocking of the parts warehouse is separately identified as a category of capital outflow. Other examples of working capital are reclamation bonds, security deposits, prepayments (rental car vouchers, airline tickets, etc.), and advances for such things as employee housing loans.

The biggest investment in working capital in a mine is the "inventory" of advance workings (e.g., advance overburden removal, floor stocks, stockpiles). Nevertheless, the mine plan typically provides for these in the mining schedule, so the cash flow associated with them is already incorporated into the operating costs of the mine.

In a cash flow calculation, working capital is treated differently than in a traditional accounting calculation. A cash flow calculation requires the flow of cash to be included exactly when it occurs. There is a direct cash outflow at the start of the project when the initial warehouse inventory is built up. This "investment" in spare parts is not tax deductible or depreciable in most jurisdictions, so it is usually tabulated in the cash flow separately. Expenditures are shown as a cash outflow when the initial inventory is built up, and there is a corresponding cash inflow when the inventory is disposed of at the end of the mine life.

Traditional accounting methods frequently report very large investments in working capital in a mine, since the objective of traditional accounting is to try to fairly represent the true operating revenues and costs associated with the production occurring only in the year reported. If a mine undertakes 2 years of advance prestripping in 1 year (by contractor, for example), then this is a doubling of cash outflow for the period and would be represented as such on the cash flow tabulation. For accounting purposes, only half of the operating expenditures would be classified as true operating expenditures, and the balance would be classified as working capital (stripping in advance) in the company accounts. In practice, this classification as working capital can often be ignored in cash flow calculations; i.e., *the expenditure should simply be classified as operating costs when it is incurred.* The only influence it might have on the project economics is the effect on taxes, and this type of operating cost is usually tax deductible in the year it is incurred anyway.

Most computer programs that are set up for cash flow analysis treat working capital just like ordinary investment capital but with a depreciation rate of zero and a salvage value equal to the purchase price.

Salvage Value The salvage value is the expected value (after cost of sale has been deducted) realized upon disposal of fixed assets at the end of their useful life. Salvage value is a cash inflow, exactly opposite to capital expenditure.

If the salvage value is different than the written-down value, then an adjustment to taxes may be necessary.

Example 5.7:

For tax purposes, the allowable depreciation for a large excavator, originally purchased at a cost of $2 million, may be a rate of 20%, straight line, over 5 years. To work out tax payable each year, the company depreciates this equipment at this rate. In reality, the excavator lasts 6 years, and at the end of this time is sold for $100,000. The salvage value of $100,000 is cash inflow, but because the written-down value for tax purposes was nil, the $100,000 of revenue from the sale is equivalent to taxable income and will be subject to tax. Alternatively, if the company could see that they were likely to be selling the equipment and that they would be likely to receive the $100,000 of income for the sale, they could choose to depreciate the excavator by a lesser amount during its life, so that at the end of its useful life it would be worth the salvage value (exactly). In this case, no tax liability would be incurred on the sale.

Timing of Cash Flows The timing of cash flows is important. Over half of all bankruptcies can be attributed wholly or largely to a miscalculation or unexpected change in the timing of cash flows. Timing is more critical in highly leveraged projects (e.g., where much of the equipment is leased) but is equally vital in three other areas, namely (1) commodities for which the prices change in cycles; (2) cases where less capital-intensive schemes are being upgraded to more capital-intensive schemes; and (3) during mine expansions.

The effect of timing on project investment strategies is a broad subject that is covered more fully later in the text. For simpler cash flow calculations, however, the following adjustments should be considered:

- Production is not sold immediately. Depending on the terms of invoicing and sale, a 2-month delay, for example, should be accommodated by tabulating just $^{10}/_{12}$ of the revenue in the year of production, with the remaining $^{2}/_{12}$ of revenue moved into the following year. (An exception to this might be the airline business, where the revenue from the sale of tickets is frequently received in advance).
- In most mines, operating costs are paid immediately. The largest operating cost item is usually labor, and payment for this cannot be delayed. Operating costs of fuel and electricity may be payable on 30-day or 60-day terms, and these elements could possibly be delayed in the cash flow analysis.
- Payment of tax is subject to the effects of timing and, if this is critical, should be adjusted for the correct time period in the cash flow. Table 5.2 shows tax occurring in the year the taxable profit is made, but frequently this liability does not come due until the following year. Many taxation authorities are now introducing progressive tax payments in the year the liability is incurred based on notional tax estimated from prior years'

income. For short-term cash flows (particularly with businesses that are growing quickly), correct modeling of the timing of tax payments is imperative.

Depreciation, Depletion, Tax Credits, and Taxation

The biggest difference between traditional accounting and cash flow analysis occurs in the treatment of depreciation, depletion, tax credits, and taxation. Cash flow analysis is concerned only with actual cash–and the only actual cash item in the preceding list is tax. Unlike with traditional accounting, the other items appear in a cash flow tabulation *only* for the purposes of helping to work out the tax payable.

Depreciation Depreciation is the allowance for wearing out of equipment. Accountants like to think that if an item of equipment ordinarily lasts 6 years, then after 3 years it "should" be worth half of the purchase price. This may or may not be true, and accounting definitions of value should not be confused with the value for decision-making purposes. In a cash flow calculation, the depreciation rate used should be the one designated by the taxation authorities. Companies may use a different figure in their accounts if they believe that the alternative method more correctly reflects the true value of equipment, but this confusion should not enter into the cash flow. In some countries, mining companies may depreciate 100% of the value of mining equipment at the start of the project *for tax purposes.* Usually depreciation is undertaken by using the straight line method (the capital cost written off in equal increments over its designated life) or by a declining balance method. In the declining balance method, the depreciation applied in any year is calculated as a fixed proportion of the written-down value at the start of the year. Some countries allow a change in depreciation method partway through equipment life.

Depletion Depletion is the equivalent to depreciation for deposit reserves. It represents the declining value of the reserve that the taxing authorities permit a company to claim as a valid deduction from taxable profits. Every country adopts different allowances for depletion in their taxation rules, and this often depends on whether or not the country allows tax deductibility of exploration expenditures and acquisition costs.

Tax Credits Tax credits are allowances permitted by the taxing authorities over and above the expenditure actually incurred.

Example 5.8:

The cost of replacing old insulation with new materials that are more energy efficient might be $500,000. This is a valid operating expense. Government incentives that allow 150% deductibility (because the government is trying to promote energy conservation) are treated as a "normal" expense of $500,000 plus a tax credit of $250,000. If the tax rate is, say, 30%, then the effect of the tax credit is to reduce the tax bill by $75,000.

Tax credits enter the cash flow calculation only as an aid to calculating the tax payable.

An increasing trend in some countries is the introduction of "negative" tax credits–valid business expenses for which the government will not allow the full tax deductibility.

Example 5.9:

Some examples of expenses that might not be allowed full tax deductibility are so-called fringe benefits and entertainment expenses. In addition, some executive salaries are deductible as a business expense only when they are less than $1 million annually. The difference between the actual expense and the deductible expense is treated as a tax credit (either positive or negative).

Taxation Taxation is the amount of money that has to be paid to the government based on the "taxable" profits. Other forms of taxation such as sales tax, goods and services tax (GST), value-added tax (VAT), payroll taxes, and state taxes are normally treated as ordinary costs of doing business. The taxable profit is the profit *as defined by the government* based on the rules just discussed; it may or may not be a fair representation of the real profit by normal accounting standards.

Applying Accounting Rules to Cash Flow Calculations

Unlike capital goods, revenues are usually received and operating expenses are usually incurred in the same production period to which they relate. Thus, there is a lot of similarity between the accounting treatment of revenues and expenses and cash flow treatment of revenues and expenses. This similarity encourages analysis of these items together–preparing the cash flow tabulation through modification of already-prepared accounts. For experienced analysts, this is the easiest way to prepare a cash flow.

The normal method of undertaking this work is to start with the profit as reported in the accounts. This after-tax profit would have had depreciation, depletion, and tax credits deducted (in order to work out the taxable income in the first instance), so the "real" after-tax cash flow will have to have these items added back. This gives rise to the frequently confusing notion that depreciation is a revenue as well as an expense. Depreciation is an expense when it is first deducted in order to calculate the tax, and it is a revenue when it is later added back to calculate the cash flow. Similar confusion arises with adjustments to working capital–some of which are deductible *for tax purposes* in the year incurred and some of which are not.

Simple cash flows such as presented in the early chapters of this text offer valuable decision support for an enormous amount of technical work. However, they are only infrequently used. One reason for this is the introduction of accounting terms into a calculation that can be reliably undertaken and better understood by using simpler constructs. For these sorts of evaluations, it is easiest to avoid the accounting approach. Depreciation, depletion, and noncash items should be separated out in the cash flow and grouped so that it is obvious they are tabulated only for the purposes of calculating the tax payable. The four rows with shaded backgrounds in Table 5.2 are included in the cash flow only for the purposes of calculating and displaying tax-related data.

Accounting rules also impact the calculation of capital values and return-on-assets calculation. These aspects are taken up in Chapter 9.

DISCOUNT FACTORS, RISK, UNCERTAINTY

Almost all decision making using time-value-of-money concepts revolves around selection of the appropriate discount rate. In most mining companies, the rate that the management applies is outside the control of the planning personnel—it is specified by the board or senior management. The elements making up the discount rate are as follows:

1. The basic interest rate applicable for zero-risk investments in the country.
2. Allowances (i.e., additions) for the cost of capital. This comprises both equity funding and debt funding.
   - Debt funding will be priced at the market's determination of risk, accounting for the track record of the company, as well as the degree of recourse the lender has to the project or other company cash flows.
   - Equity funding costs will be a function of the stock market valuation of the company, which will again be a function of the track record of the company and the amount of debt funding.
3. The cost of capital is dependent on the relative mix of the debt and equity. Since debt funding is usually of lower cost than equity funding, higher debt:equity ratios may mean lower costs of capital but higher financial risk. (The simple division of funding between "debt" and "equity" is increasingly being blurred. Debt that is convertible to equity, for example, and preference shares that have prioritized claims on profit distributions are just two financial instruments within a full spectrum of options available to corporate finance departments.)
4. Allowances for the finance risk. Lenders have first recourse to the project's surplus cash flows, and the "premium" on financing costs will depend on the lenders' perception of the likelihood of loan repayment. Even if the financiers are assured of payment (by a mortgage over some other company property, for example), this raises the cost of finance because it inhibits the company's own ability to raise capital elsewhere.
5. Allowances for the technical risk. Geotechnical characteristics may cause losses of production or higher costs of production. Grades of the orebody may not turn out to be as predicted. The discount rate must also reflect the fact that the company must commit to apply (and get some premium for) its technology to manage a mine for a long time and actually bring the cash flows to fruition. Some of the technical risk may not apply to the investing partner in a mining venture as fully as for the operating partner, so investors who are passive participants in the venture may be able to adopt a lower discount rate for their investment in the same project.
6. Allowances for the fact that the company is more "locked in" to its investment. Money invested in government securities, for example, can be retrieved immediately on the open market. Money invested in mining projects cannot be retrieved as easily, and the difficulty in exiting the project requires a premium not applicable to many other investments and (perhaps) not applicable to the nonoperating stakeholders in the project.

7. Allowances for the fact that, for a company to stay in business in the long term, the returns for this project must also cover the costs expended in assessment of projects to potentially replace this one when it is exhausted. It is not sufficient in the long run for the project just to cover its own assessment and acquisition costs.

Overriding the preceding considerations is the problem of limited funds. This applies to all companies. If funds limitation is critical (which it would be for all but the smallest projects), then the rate must be at least as high as possible rates obtainable from alternative investments.

INFLATION AND CONSTANT MONEY CALCULATIONS

Throughout the 1970s and 1980s, in most of the western world, inflation was an all-pervading influence on decision making. It continues to be a strong influence in the 2000s in many other parts of the world. The problem can be partly alleviated by preparing the complete cash flow in a low-inflation world currency (such as U.S. dollars)—even if the project is located in some high-inflation, third world country, for instance.

The effects of inflation must be separated out of the calculation if meaningful decisions *pertaining just to the business proposition* are to be made. A business proposition that returns 15% on capital invested in a 15% inflationary environment is not a business at all. The money can be put into noninflating foreign currencies or solid assets for which the value is increasing with the general rate of inflation, with less risk and the same effective return.

For simple cases, inflation is ignored. Does this mean that if the rate of return is 15% and there is 5% inflation that the real return is just 10%? No. A discounted cash flow analysis that ignores inflation is effectively assuming that revenues and costs escalate at the same amount and cancel each other out. Despite a strong emphasis on inflation-based analysis in many companies, the reality is that for most calculations the simple, noninflationary scenario yields quite satisfactory results. The precision (so-called) of inflation-adjusted cash flows probably makes less difference to the result than many other factors that are not even shown in the calculation!

If inflation is considered important, and particularly if some costs or revenues are changing at different rates than others, then an inflation-adjusted discounted cash flow must be prepared. This form of cash flow adds additional steps: selling prices are escalated at their estimated rate; and costs, including replacement costs of capital, are escalated at their estimated rates. The cash flow analysis is undertaken in the same manner as in Table 5.2. The resultant cash flows in each year are both future currency *and* inflated currency.

TABLE 5.3 Sample inflation-adjusted discounted cash flow

	Year					
	0	1	2	3	4	5
Production, *oz*		30,000	50,000	50,000	50,000	45,000
Unit selling price, starting at $500/oz, *thousand $*		545	594	648	706	769
Operating revenue, *thousand $*		**16,350**	**29,703**	**32,376**	**35,290**	**34,619**
Operating expenses:						
Unescalated operating expenses, *thousand $*		10,598	17,762	19,339	21,073	20,882
Labor component: 55.0%, *thousand $*		5,829	9,769	10,636	11,590	11,485
Other component: 45.0%, *thousand $*		4,769	7,993	8,702	9,483	9,397
Escalation factor, labor component		*1.110*	*1.232*	*1.368*	*1.518*	*1.685*
Escalation factor, other component		*1.090*	*1.188*	*1.295*	*1.412*	*1.539*
Total operating expenses (escalated), thousand $		*11,668*	*21,533*	*25,816*	*30,980*	*33,811*
Operating profit, *thousand $*		**4,682**	**8,169**	**6,559**	**4,310**	**808**
Capital expenditure, *thousand $*	15,000					
Tax depreciation, at declining balance rate of 27.5%, *thousand $*		4,125	2,991	2,168	1,572	1,140
Written-down value for tax purposes, *thousand $*		10,875	7,884	5,716	4,144	3,005
Unescalated salvage value, *thousand $*						3,005
Salvage value (escalated), *thousand $*						4,623
Profit as assessed for tax purposes, *thousand $*		557	5,178	4,391	2,738	1,287
Income tax payable at rate of 35%, thousand $		195	1,812	1,537	958	450
Cash flow in dollar-of-the-day terms, *thousand $*	(15,000)	4,487	6,357	5,022	3,351	4,980
Inflation correction factor at 10.0%	*1.000*	*0.909*	*0.826*	*0.751*	*0.683*	*0.621*
Cash flow in constant dollar terms, *thousand $*	(15,000)	4,079	5,253	3,773	2,289	3,092
Discount factor at rate ***8.2%***	*1.000*	*0.924*	*0.855*	*0.790*	*0.730*	*0.675*
Discounted cash flow, *thousand $*	(15,000)	3,771	4,489	2,981	1,672	2,088
Net present value, *thousand $*	**0**					

Note: Numbers in parentheses indicate negative values.

Table 5.3 shows the same cash flow as in Table 5.2, with the following adjustments for differential annual rates of inflation:

General inflation rate	10.0%
Product selling price	9.0%
Labor component of operating costs	11.0%
Other component of operating costs	9.0%
Capital equipment	9.0%

In Table 5.2, the data were deliberately chosen to yield a 15% internal rate of return at a gold price of $500/oz. In Table 5.3, the internal rate of return (i.e., the rate that equates cash inflows and outflows to yield an NPV of zero) is 8.2%. If a discount rate of 15% had been applied to the constant-dollar cash flow of Table 5.3, the net present value would have been calculated to be −$2,153,000.

The two discounting steps at the bottom of Table 5.3 are required to determine the inflation-adjusted, or real, return on investment. The first step is to turn the inflated cash flows into constant money values by discounting them at the rate of general inflation. Values in each year then represent future currency only, but currency with constant purchasing power. It is like saying if a dollar will buy an apple today, a dollar 3 years hence will still buy an apple 3 years hence. However, the receipt of the dollar (or the apple) 3 years hence is still worth less than the receipt of a dollar (or an apple) today. To determine the internal rate of return, cash flows are discounted again to find the rate that sets the net present value to zero.

Table 5.3 yielded a different (lower) real rate of return than Table 5.2 because some of the costs (labor) were escalating faster than the general rate of inflation, while revenues were escalating at rates less than the rate of inflation. Even small differential rates can make substantial differences in the return.

One problem with this approach is the imprecision in estimating the differential rates of inflation for the different components of the cash flow. In this case, many practitioners use one rate for all items in the cash flow (at least as a first estimate). Does this make sense? If all amounts are inflated up and then deflated down back to a constant money equivalent, then why do it at all?

The answer lies in the tax calculation. To determine the taxable income, the rules usually allow depreciation deductions based only on the historical cost of equipment. The result is that, over time, the total amount of tax paid increases as a proportion of the revenues. The real rate of return declines.

Even in the simple case shown in Table 5.3, with all rates of escalation set at 10%, the impact on the project, compared to Table 5.2, would be (1) a reduction in net present value by $749,000, or (2) a real internal rate of return that reduces to 12.8%, or (3) a gold price that would have to be $507.80 in order to maintain the real (required) return of 15%. These changes are entirely due to the additional tax payable. The tax payable increases from 3.47% of revenues in constant dollar present value terms from Table 5.2 to 4.96% of revenues in the constant 10% inflate/deflate case–an effective 43% increase in tax. (This result demonstrates why governments have only limited interest in reducing inflation. Of course, the benefit the government sees from higher taxes is applicable only to projects that remain viable, and the uncertainties and lower returns caused by inflation ultimately reduce the number of mines that fall into this category.)

With relatively low inflation (less than 4%) in most of the developed world, the noninflationary model similar to Table 5.2 should be followed for simple cases. In more detailed analyses, the inflate/deflate model similar to Table 5.3 should be followed. Some organizations even exploit this effect where different escalation rates apply to different elements of the cash flow. Example 5.10 illustrates.

Example 5.10:

A contract earthmoving project might have provisions for escalation of operating supplies but only partial escalation on capital replacements. In addition, the general inflation rate may be quite different than the inflation rate for the kinds of goods and services used by the resources industry. In contract negotiations an apparently low initial bid price can soon favor the supplier through escalation provisions exceeding the true rate of price increase of the underlying commodity.

The earlier warning about differential rates of inflation bears repeating. Inflation-adjusted cash flows using differential rates of escalation for inputs and products can quickly lead to erroneous results unless very carefully applied. Over a 20-year project life, a small difference in projected escalation rates can result in very large relative prices in the underlying commodities. If the relative price of different inputs (e.g., fuel oil, labor, electricity) or mine outputs does change by such large amounts, then the relative economics of mining would change. If the mining method changed as a result of inflation, then the cash flow analysis would no longer be valid.

DISCOUNTED CASH FLOW RANKING CRITERIA

How does a company decide whether a project should proceed or not, or whether one project is better than another project? The two most common measures are based on

- net present value
- return on investment (or internal rate of return)

In either case, higher values are preferable. If a company's cost of capital is 10% and the net present value is $100 million after discounting the cash flow at this rate, then the value of the project to the company is $100 million. On the basis of this criterion, an alternative project with a net present value of $120 million would be a better project.

Alternatively, the discount rate can be selected so that the cash outflows exactly balance the cash inflows (in present value terms). At this rate the net present value will be zero. This rate is termed the internal rate of return. On the basis of this criterion, a project with an internal rate of return of 19.5% would be better than a project that has an 18.5% IRR.

Either criterion can be used; however, the methods do not always yield the same answer. A small project with a very high return might have a lower net present value than a larger project with a lower return.

The difficulty in understanding which criterion is best stems from the definition of what constitutes a "project." Recall from Chapter 4 that choices are always between one course of action and some other course of action that will have to be forsaken, but the definition of each alternative must encompass the full scope of the differences between the two. If a company can afford either but not both projects and proceeds with the smaller one, what will the company do with the capital still available? If this money is otherwise to be left in the bank, then the return on these bank funds must be taken into consideration in a comparison of the smaller project with the larger project. (Most books on corporate finance include discussion on the relative merits of the different investment criteria and why one or another method is superior. For a comprehensive discussion, see, e.g., Brealey and Myers [2003].)

In practice, even after detailed study, the characteristics of each investment opportunity are seldom understood very well at the start. Just because a project has a return exceeding the cost of capital does not mean it should necessarily proceed, even if there are no better projects immediately available. Indeed, the cost-of-capital guideline is only a very blunt instrument for project selection. Typically the hurdle rate–the minimum internal rate of return–insisted upon by large corporations exceeds the cost of capital by factors of three or four (Pindyck and Solimano 1993). The final decision cannot be quantified as precisely as standard discounted cash flow procedures suggest. Nevertheless, with all its failings, the procedure still accurately characterizes the approach adopted by most large organizations throughout the world in undertaking this kind of assessment.

Stochastic models and various decision analysis tools (themselves often variations of the DCF technique) are frequently used to address particular areas of concern. Some of these extensions are discussed in Chapter 12 and later chapters.

DISCOUNTED AVERAGE COST

The capital recovery function (see Equation 5.3 and Example 5.3) defined the relationship among a present value, regular future cash flows, and the interest (or discount) rate. By using this formula for any one interest rate, one can calculate a constant stream of cash flows from any present value. Alternatively, given a known constant stream of cash flows, one can derive the equivalent present value.

The same relationship can be calculated by using a discounted cash flow table. Indeed, with some exceptions, any of these methods can be used with two available or assumed inputs to calculate the third:

For any:	*Coupled with:*	*One can calculate:*
Selling price of the product (stream of cash flows)	Discount rate set to the cost of capital	Net present value
Selling price of the product (stream of cash flows)	Net present value set to zero	Internal rate of return
Hurdle rate, or benchmark rate of return	Net present value set to zero	Selling price that must be obtained

Tables 5.2 and 5.3 have already demonstrated this sort of calculation for net present value and internal rate of return.

Unlike the capital recovery formula, which allows a simple algebraic expression, a discounted cash flow table yields these results only after trial and error or after some iterative procedure. (The net present value calculation yields a unique result. However, in some cases there may be more than one rate of return that will yield a net present value of zero; i.e., it is possible for there to be more than one internal rate of return.) Nevertheless, most spreadsheets allow these procedures to be readily undertaken, and the alternative ways of presenting results yield valuable insights for decision making. It is the third of the tasks listed above—determining the selling price—that is the subject of this section.

The standard DCF technique *starts* with the expected selling price. But what if there is no market price? A competitive bid for long-term supply of some commodity—for example, a long-term coal supply contract to a captive power station—requires the bidder to *nominate* the price.

In this case, the "market" price is the estimated price that the lowest-priced competitor is likely to bid—an estimate arrived at by analysis of the competitor's estimated cash flow. Alternatively, a company preparing a bid could nominate a price that, if the bid were successful, would yield some desired return on investment. A similar situation exists for businesses pricing new products that are not yet available in the market.

Most capital investment decisions on mine sites will not be related to an actual market-saleable product. If a firm is comparing two alternatives for reclamation, usually *both* alternatives produce the same output (acres of reclaimed land), which the firm doesn't sell and which doesn't relate to any change in mine output. Yet the firm still wants to ensure that its investment in the equipment yields an appropriate return and that the "cost" that it applies to the reclamation reflects this return. For this sort of calculation, the DCF method uses an assumed price, or internal price.

Internal prices are determined through a discounted average cost calculation. The discounted average cost is defined as the price per unit of production you would have to pay someone else *with the same investment criteria as yourself* to have the production undertaken.

Internal Pricing

For economic decision making, internal pricing finds its most important application with activity-based costing. Unlike traditional accounting, which has difficulty in the allocation of capital across different parts of the mine, the discounted average cost calculation readily allows this calculation. It also includes tax in its derived cost of production. One can subdivide the mine into as many sub-businesses as warranted, allocating capital to each, and then work out internal prices for production from each sub-business.

The internal pricing calculation starts with the investment and planned operating cost structure associated with the sub-business activity and *determines* what selling price per unit of output yields the required overall return. The internal price (transfer price), or discounted average cost of production, is

TABLE 5.4 Example of open pit coal mine subdivided into six sub-businesses

	Description of Activity	"Product" Produced
1.	Waste drilling and blasting: preparation of overburden prior to excavation	Cubic meters of broken ground
2.	Waste excavation and rehabilitation	First product: coal exposed—expressed in square meters of surface area of coal (or in situ quantity of coal). Second product: area of rehabilitated land.
3.	Coal excavation and hauling	Tons of raw coal placed in a hopper
4.	Coal processing and load-out	Tons of saleable coal product placed on a train. The output from this sub-business is sold in the external market.
5.	Maintenance	Equipment maintenance and servicing—charged on hourly rates or fixed quotation with availability guarantees
6.	Coordination, planning, marketing, and financing	Efficiency in capital usage, entrepreneurial profit opportunities, services

applied to the product of the sub-business. Products of individual sub-businesses may bear no relationship to the products of the firm as a whole.

The formulation is best illustrated with an example of an open pit coal mine subdivided into six sub-businesses (Table 5.4). As the table shows, sub-businesses 1 through 4 follow the production chain. Sub-businesses 5 and 6 are service businesses. The definition of the "product" must be unambiguous and unique for a given sub-business, but since the product is being sold only internally, it need not be market saleable. For instance, "square meters of exposed coal" in sub-business 2 is a well-defined output from this entity, yet as a commodity it is valuable only to the next sub-business in the production chain.

A Sample Discounted Average Cost Calculation

As a simple illustration, assume that one sub-business consists of a single large crawler dozer used for earthmoving. Capital, operating cost, and production characteristics of the dozer over its 4-year life are set out in Table 5.5. In this example output declines by 0.25% each 1,000 hours of machine life, and fuel and lube costs, repair part costs, and maintenance labor costs increase as the machine ages.

The discounted average cost is the price per unit of production you would have to pay someone else to have the production undertaken. It must incorporate the taxation effects associated with operating "profits" and depreciation. For this example, straight line depreciation has been assumed, set to an annual amount that will lead to write-down of the book value to the salvage value at the end of the machine life. The "required" after-tax return on investment is 15%. (The required return of 15% for this sub-business implies that, for consistency, the whole project of which this sub-business is a part has an expected return equal to this percentage.)

TABLE 5.5 Base data for discounted average cost calculation

Operating or Production Component	Base Costs and Rates	Change Each 1,000 hours
Hourly production	1,000 units/operating hour	–0.25%
Annual usage	5,000 operating hours	No change
Initial capital cost	$750,000	—
Salvage value at end of 4 years	$ 75,000	—
Fuel and lube	$18.00/operating hour	+0.50%
Tracks and wear items	$ 4.00/operating hour	No change
Repair parts	$31.00/operating hour	+1.00%
Operating labor	$45.00/operating hour	No change
Maintenance labor	$21.00/operating hour	+1.00%
Tax rate	35.00%	—
Discount factor (required return on investment)	15.00%	—

Table 5.6 sets out the discounted cash flow. It has been prepared by using an end-of-year convention; it shows production and revenue in the first rows of the table, followed by capital expenditures and depreciation calculations and then operating costs (including the tax payable). The final rows of the table show the cash flows year by year, as well as a discount factor to turn these into present value terms.

The only point of differentiation between the cash flow in Table 5.6 and a regular discounted cash flow analysis is the way that the unit revenue (shown on the sixth line of Table 5.6) is calculated. In a standard analysis, this value is externally determined. In a discounted average cost calculation, it is internally determined by using an iterative procedure to equate the net present value at the bottom of the table to zero. The *objective* of the calculation is to determine the discounted average cost of production per unit. (The unit revenue is usually calculated with the "goalseek" or "backsolver" feature available on most spreadsheets.) This value is the transfer price for "sale" of the output from the dozer that, when applied to the expected production, will yield the required 15% return on investment for that machine.

The discounted average cost calculation is a powerful tool for faithfully comparing alternatives with different capital and operating cost characteristics and for comparing "internal" prices (cost of production) with outsourced alternative prices. Comprehensive examples using this form of discounted cash flow analysis are included in Chapters 7 and 10.

How does one apply this procedure for the internal pricing of the six activities described in Table 5.5? The procedure requires a whole-mine cash flow to be completed first and then all of the capital in the mine to be allocated or apportioned to activities.

The whole-mine cash flow establishes the project expected return and the market-based reference for the individual activities. For budgeting purposes, each activity must also achieve this return, based on its "product" sold internally.

TABLE 5.6 Discounted average cost calculation

	Year				
	0	1	2	3	4
Machine hours, start of year		0	5,000	10,000	15,000
Age of machine (total machine hours), mid-year		2,500	7,500	12,500	17,500
Machine hours, end of year		5,000	10,000	15,000	20,000
Production rate this year, *units/hour*		993.8	981.4	969.2	957.1
Production this year, *units*		4,968,809	4,907,008	4,845,977	4,785,704
Gross revenues this year ***at $0.1871 per unit of production,*** $		929,792	918,228	906,807	895,529
Capital expenditure, $	750,000				
Trade-in or salvage value, $					75,000
"Book" value, start of year, $		750,000	581,250	412,500	243,750
"Book" value, end of year, $		581,250	412,500	243,750	75,000
Depreciation, $		168,750	168,750	168,750	168,750
Operating expenses					
Fuel and lube, $		91,129	93,430	95,790	98,208
Tracks and wear items, $		20,000	20,000	20,000	20,000
Repair parts, $		158,904	167,010	175,529	184,483
Operating labor, $		225,000	225,000	225,000	225,000
Maintenance labor, $		107,645	113,136	118,907	124,972
Total operating costs, $		*602,678*	*618,576*	*635,225*	*652,663*
Unit operating costs, $		*0.121*	*0.126*	*0.131*	*0.136*
Operating profit, $		327,114	299,652	271,582	242,865
Profit for tax purposes, $		158,364	130,902	102,832	74,115
Tax payable, $		55,428	45,816	35,991	25,940
Cash flow, $	(750,000)	271,687	253,836	235,591	291,925
Discount factor	*1.000*	*0.870*	*0.756*	*0.658*	*0.572*
Present value of cash flow, $	(750,000)	236,249	191,937	154,905	166,909
Net present value, $	0				

Discounted average cost of production = $0.1871/unit (from the unit revenue shown on line 6 of table).

Note: All numbers shown in parentheses in table indicate negative values.

Thus, the capital and operating costs apportioned to waste drilling and blasting determine the discounted average cost of "production" for broken ground. This "purchased" input, plus the apportioned capital and operating costs applied to the next activity, determine the discounted average cost of production from this activity, and so on. With consistent apportionment, transfer pricing through the chain of production will be consistent with expected returns from the whole-mine cash flow.

Activity-based budgeting after this model allows practitioners operating within their own narrow areas of expertise to concentrate on efficiency as defined by their own rules but benchmarked and constrained according to the whole-mine return on investment.

CHAPTER 6

Systematic Approach to Exploration Expenditure

How well proven does an orebody have to be before mining can start? Can potential reserves in the ground be used to justify exploration expenditures? If a blocked-out reserve (a potential mine) is clearly uneconomic, how can it be "worth" anything?

These questions, important ones in the economic assessment of mineral deposits, are logically resolvable with the same economic tools used for evaluation of viable mines and for making production decisions. This chapter demonstrates that even the uncertain economics of finding and proving up reserves are no less constrained by economics than are the more mechanical aspects of mining decision making.

SYSTEMATIC APPROACH TO EXPLORATION EXPENDITURES

As with all other aspects of mining economics, the economic assessment of mineral deposits is subject to two powerful influences (among others), namely, the time value of money and the economics of information. It is the second of these—the extra value from the extra information after extra drilling—that is the major focus of attention in this chapter.

Chapter 3 set out a systematic procedure for evaluating mining projects and potential projects. One objective of the procedure was to maximize the value from any exploration and evaluation budget. This objective was partly achieved by minimizing expenditure in areas that are clearly uneconomic, particularly if they are easily identified early in the process. Under this procedure, the economic assessment of mineral deposits starts before the first drillhole is put in the ground and continues until the mine is finally closed.

Data are always imperfect, and answers are always subject to uncertainty. Each evaluation can do no better than to use the best data available, however imprecise; a lack of information per se is not sufficient reason to delay economic assessment. Each assessment yields two results:

- the estimate of expected value (return on investment, cost of production, etc.)
- an estimate of the reliability of the results

Expected values, as well as the estimated reliability of these values, determine the need for more detailed evaluation. Decisions regarding expenditure on additional information are themselves internal to the evaluation process.

The end purpose of exploration drilling is to find and prove up mining propositions. The starting point—prior to exploration—should outline the characteristics of some *target* mining proposition. A limiting condition—the best deposit that could be hoped for—should be assumed and its economics examined. If the "best" deposit is not economical, then there is little hope that lesser ones will be.

An economic assessment of an assumed orebody is an important guideline for exploration. At minimum, it helps establish initial exploration priorities between, say, lower-grade shallow ore and higher-grade deeper ore. It also establishes priorities between reserve extensions that allow mine expansion and reserves extensions that increase mine life.

If the "best" deposit is not viable, it does not mean that exploration should not go ahead, but it does mean that exploration must be justified on some basis other than finding a potential mine nearby.

Once the technical characteristics of an orebody start to be understood, an economic assessment must be carried out. If nothing else, this first assessment ensures that future exploration budgets are productively spent. This first assessment—part of the strategic planning phase—should be started as soon as possible within the exploration phase and certainly prior to any detailed planning. The logic for this early assessment is constrained by three steps:

1. For the detailed planning (project approval), detailed exploration is normally warranted only for the first few years of mining. Later in this chapter, a simple economic analysis validating this constraint is given.
2. The parts of the deposit that will be exploited during the first few years of mining are quite ill defined until the strategic planning phase is complete. Until this stage is complete, the direction of mining, the development strategy, timing, and many other likely constraints on development are also very poorly understood.
3. It is normally too costly to drill out in detail *all* of the possible starting points of the mine or to preassess all of the likely constraints. Even if the cost of exploration is small compared to the value of the deposit, constraints and limits that apply to one method may be unimportant for other methods or mutually exclusive to each other. At the very least, much effort is wasted. *Detailed* exploration work cannot be commenced until after the strategic planning defines these guidelines.

The initial economic assessment of a mineral deposit is therefore undertaken at the point where the deposit is understood only in broad terms. The objective is not to come up with a particular mine plan (although the preparation or attempted preparation of a plan is frequently the most convenient way to prepare such an assessment). The objective is to determine the economics of a *likely* mine plan, assuming such a plan is feasible. Future evaluation work is then directed at proving up the components necessary to ensure this feasibility.

This chapter examines the economic assessment of mineral deposits along three fronts. First, the economics of exploration are examined in broad terms. The value of reserves in the ground is examined. This valuation provides guidelines for how to establish additional exploration priorities and how additional exploration can be justified.

Next, the text discusses three techniques aimed at understanding the economics of identified deposits that are not yet well enough understood for a robust mine plan to be developed. It follows convention in that many initial assessments focus primarily on operating costs and frequently ignore time-value-of-money and taxation impacts.

Finally, the last section of this chapter aims at understanding how fixed costs, variable costs, and the time value of money influence decisions as to whether particular parts of the deposit are viable or not. This section applies equally to exploration at the early stages of a project or to alternative mine development schedules within an operating mine.

UNDEVELOPED DEPOSITS

The valuation of undeveloped deposits is an important part of the economic assessment spectrum, as well as one of the most critical and sensitive issues concerning the worth of exploration and mining companies. There is a wide consensus on how these sorts of valuations should be conducted (i.e., on the basis of the present value of future cash flows), yet huge differences in valuation can commonly be observed. The reasons for this are readily explainable and arise from the differences in valuations in three areas:

1. The use of average values. An "average" valuation (per ounce of gold in the ground, for example) includes some reserves that will be mined next year and some that won't be mined for another 20 years. In present-day economic terms, these are vastly different reserves.
2. The degree of confidence. Reserves are worth the (present value of the) difference between their cost of production and the selling price. There may be a lot of volatility and uncertainty in the selling price. There may be a lot of uncertainty about the cost of production. If all other factors are equal, more uncertainties mean lower value of the reserves.
3. Timing. Reserves are worth more if they are brought into production sooner. Their value is also dependent on the impact that additional production has on costs or profitability elsewhere.

Runge (1994) documents a number of examples, parts of which are summarized in the following subsection.

TABLE 6.1 Unit value of reserves based on the life of the reserves

	Year				
	1	2	3	4	5
3-year-life reserves:					
Production, *tpy*	1,000,000	1,000,000	1,000,000		
Cash flow, *$*	12,000,000	12,000,000	12,000,000		
Discount factor at rate of 15%	*0.870*	*0.756*	*0.658*		
Present value of cash flow, *$*	10,430,000	9,070,000	7,890,000		
Total net present value and average value: $27,400,000 for 3,000,000 t of reserves = $9.13/t					
4-year-life reserves:					
Production, *tpy*	1,000,000	1,000,000	1,000,000	1,000,000	
Cash flow, *$*	12,000,000	12,000,000	12,000,000	12,000,000	
Discount factor at rate of 15%	*0.870*	*0.756*	*0.658*	*0.572*	
Present value of cash flow, *$*	10,430,000	9,070,000	7,890,000	6,860,000	
Total net present value and average value: $34,260,000 for 4,000,000 t of reserves = $8.56/t					
5-year-life reserves					
Production, *tpy*	1,000,000	1,000,000	1,000,000	1,000,000	1,000,000
Cash flow, *$*	12,000,000	12,000,000	12,000,000	12,000,000	12,000,000
Discount factor at 15%	*0.870*	*0.756*	*0.658*	*0.572*	*0.497*
Present value of cash flow, *$*	10,430,000	9,070,000	7,890,000	6,860,000	5,970,000
Total net present value and average value: $40,230,000 for 5,000,000 t of reserves = $8.05/t					

Reserves in the Ground: An Example

All estimates of the value of reserves in the ground begin with an estimate of the difference between the likely cost of production–including capital costs for any long-term valuation–and the likely selling price. This difference represents the likely future cash flows to repay the cost of finding and/or acquiring the reserves. The present value of the likely future cash flows is calculated by discounting at the company's opportunity cost of capital. By convention, reserve valuations are normally presented as a dollar value per unit of reserves in the ground.

Table 6.1 shows a simple tabulation of three hypothetical cases representing a deposit with reserves extending for 3, 4, or 5 years. In each case, production is 1 million tpy, selling price is $20/t, and total costs of production (including capital costs) are $8/t. If the time value of money were irrelevant, all of the reserves would be valued at $12/t of ore in the ground.

Table 6.1 has three cases with identical costs of production, yet it illustrates important distinctions for reserve valuation:

1. Reserves that can be exploited sooner are worth more on a per unit basis. The 3-year-life reserves are valued at $9.13/t, whereas the 5-year-life reserves are valued at only $8.05/t *on average*.
2. An extension of reserves that extends the mine life is not as valuable as an extension of reserves that allows an increased production rate for the same mine life. A 25% increase in reserves from the 4-year-life case to the 5-year-life

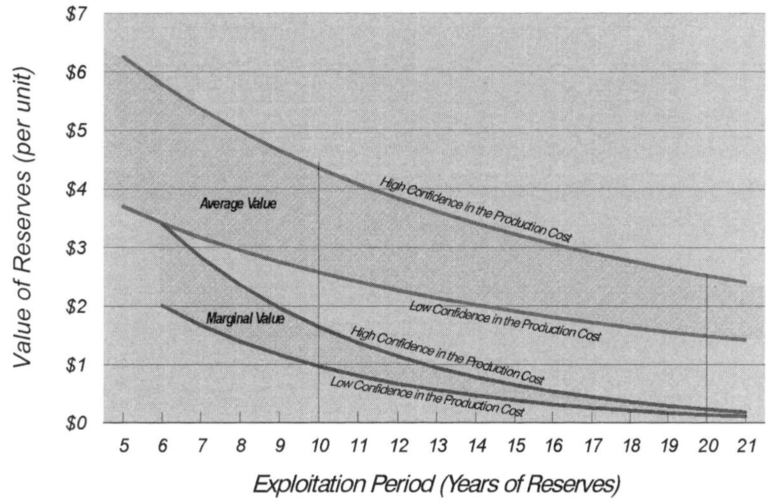

Note: Figure derived from Figure 8 in Runge (1994).

FIGURE 6.1 Value of reserves, including uncertainty, for sample simulation model

case has a marginal value of $5.97/t. If the same reserve extension allowed higher production in year 4 instead of an allowing an extra year of production—even if the costs of production were unchanged—the reserves would be valued 43% more ($8.56/t).

The trivial example in Table 6.1 is inadequate for real-life applications, which are influenced by uncertainties in reserve definition, reliability or unreliability in production cost and selling price estimates, and tax considerations. Decision makers will have an idea of the *expected* costs of finding reserves and the *expected* costs of production. However, exploration targets will be selected on a probability basis—the probability that the value of new reserves will exceed the cost of finding them. Decisions under uncertainty are made according to slightly different criteria than used for decisions that involve no uncertainty, but outcomes for the purposes of this discussion are similar.

More robust valuations require modeling of the variability and the influence this variability has on assessed value. A sample simulation model extending the data in Table 6.1 is used in the balance of this section for illustrative purposes. In this model, a tax rate of 35% has been applied, depreciation (depletion) of reserves for tax purposes over the life of the reserve has been accounted for, and production cost uncertainty has been characterized by a normal distribution. (Similar analyses can be undertaken using "add-in" software with most common spreadsheets or with the built-in random number and statistical functions of the spreadsheet software itself.) No variability was applied to the selling price. An opportunity cost of capital rate of 15% was used for this evaluation. Reserves associated with a project life varying from 5 to 20 years are valued as shown in Figure 6.1.

In Figure 6.1, the "average" value is the value under the assumption that reserves can be exploited throughout the period. Suppose a company currently has 10 years of reserves. If it finds 10% more reserves *and* expands production by 10%, then the "average value" set of curves should be used. The "marginal value" set of curves, on the other hand, should be used if the reserves are assumed to *extend* the period of exploitation—in this case, the 10% extra reserves extend mine life out to 11 years.

Results can be deduced from Figure 6.1 in a number of ways:

1. A reserve that has economics similar to what is shown in Figure 6.1 and that can be exploited over a 10-year period beginning immediately, with a high confidence in the production cost, has an average value of $4.35/t—or about 36% of the $12.00 current-day cash flow per unit shown in Table 6.1.
2. If the market has only low confidence in the technology used to exploit the deposit, it places very little value on increases in reserves. The present value of 20 years of reserves at low confidence levels is much lower than the value of 10 years of reserves at high confidence levels. In this situation, valuations based on reserves in the ground are less reliable than valuations that focus on precision in production cost estimates. For owners of this type of undeveloped resource, the management focus must be on refinement of the technology for production, not reserve extensions.
3. Even with high production costs, if there is also a high confidence in the production technology, then reserves in the ground are very valuable *if they can be exploited quickly.*

There has been a tendency in many mining texts to consider *reserves* as a technical term, independent of economics. This is understandable, given a history of rich deposits for which reserve valuations were much less sensitive to the economics of exploitation. At the same time, statutory reporting—perhaps the most common reason for the preparation of reserve estimates—is concerned only with physical quantities, not economically viable quantities. In the future, the *economics* of the reserve definition will be much more vital. Apart from the obvious selling price and production cost assumptions, critical factors in valuing reserves (and justifying additional exploration) include a firm understanding as to the production rate, likely variability in production rates, production cost confidence, and timing of cash flows.

Mining Reserves: An Example

In Chapter 2 a simple calculation was used to illustrate why mine reserves (or indeed, the quoted world reserves) differed among minerals. Reserves of some commodities (e.g., silver) were small in terms of years of life, while other commodities (e.g., bauxite) had "proven" reserves of over 200 years at current rates of annual production. Economics, rather than the physical amount of material in the ground, was shown to be the major reason for this.

This section sets out a more realistic example of the optimum level of reserves in a productive mining application.

TABLE 6.2 Two-year reserves (with fixed components of mining cost)

	Year		
	0	1	2
Annual production, *t*		1,000	1,000
Marginal revenue at selling price of $40.00/t, $		*40,000*	*40,000*
Access development cost, $	25,000		
Direct operating costs at $25/t, $		25,000	25,000
Marginal cash flow, $	(25,000)	15,000	15,000
Discount factor at rate of 15.0%	*1.0000*	*0.8696*	*0.7561*
Present value of marginal cash flow, $	(25,000)	13,043	11,342
Net present value, $	*(614)*		

Note: Numbers in parentheses indicate negative values.

Consider, for example, almost any reserve block forming part of an active mine. A new or extended orebody in an underground mine will require specific access development for which the cost is independent of the amount of reserves in the block. Similarly, open pit mines incur access development costs and development costs for the boxcut. For tax purposes these costs are operating costs. Many internal company accounts also classify them as operating costs,* but they are nevertheless costs of a "fixed" nature independent of the reserves exploitable via the development. The *decision* to mine this reserve block and to expend these fixed costs is a function of the amount of *expected* reserves in the block.

Table 6.2 shows a case similar to the example in Chapter 2, where an already delineated reserve block is adequate for 2 years of production. Based on the data in Table 6.2, the cost of developing access into the reserve area for only 2 years of mining is not viable. If additional reserves (accessible from the same development) could be found, the development *may* be viable. If additional reserves exist, suppose they can be proven up for an expected expenditure of $5/t, or $5,000 expended at year 0. Is this an economical strategy to follow? Table 6.3 shows the same data as Table 6.2 but with the addition of production for a third year and recalculation of the net present value.

In Table 6.3 the NPV has increased by the present value of the marginal cash flow. Subtracting the expected $5,000 of extra exploration necessary to demonstrate this additional reserve still yields a positive NPV. Additional exploration is viable, and exploitation of the reserve is now economic.

If it is economic to extend reserves from 2 years of life to 3 years of life, then is it also economic to extend it for a further year? On a narrow interpretation,

* Significant costs of this nature are normally treated as capital for accounting purposes if they are recognized as such. An ideal accounting system would allocate costs according to the expected reserves and make appropriate adjustments when the reserve quantity changes in operation (reserve depletion or extension). The example in Table 6.2 is valid regardless of how these costs are treated in the accounts.

the answer is yes. It is economic to prove up reserves until the marginal cost of proving them up equates to the marginal return. The marginal return is the present value of the marginal cash flow. Up to 7 years of reserves can be drilled out in advance. In this case, the NPV (after exploration cost) is maximized at $12,406.

However, this interpretation fails one of the important tests described in Chapter 4. It confuses the cost of the project with the cost of the decision. The *decision* to exploit the reserves is correct if the return from following this course of action is higher than the return from any alternative course of action. Extension of reserves from the 2-year life case to the 3-year life case—assuming the reserves are actually proven—changes the choice. Without the additional drilling, according to the investment rules, the reserves would be abandoned. With additional drilling the reserves can be mined. *Even more* drilling does not change this decision. Once the decision to mine the reserves is made, there is always the possibility of drilling out additional reserves later. Thus, the alternative to drilling 7 years in advance is to drill no more than 3 years in advance initially and then drill out subsequent reserves immediately prior to extraction. Delaying the drilling does not change the decision, but it does delay the expense and therefore improves the cash flow.

INITIAL ASSESSMENTS

Once a deposit has been identified, initial economic assessments can commence. Techniques for initial economic assessments have been developed for different types of deposits; however, the primary economic objectives are similar. Three techniques used as starting points for this are cutoff-grade analysis, pit optimization, and cost ranking:

1. Cutoff-grade analysis is used in mines where ore boundaries are defined primarily by economic criteria, e.g., porphyry copper deposits, polymetallic ore bodies, most underground (bulk) mining projects.
2. Pit optimization is used for most open pit hard-rock mines.

TABLE 6.3 Three-year reserves (with additional exploration costs)

	Year			
	0	1	2	3
Annual production, *t*		1,000	1,000	1,000
Marginal revenue at selling price of $40.00/t, $		*40,000*	*40,000*	*40,000*
Extra exploration cost, $	5,000			
Access development cost, $	25,000			
Direct operating costs at $25/t, $		25,000	25,000	25,000
Marginal cash flow, $	(30,000)	15,000	15,000	15,000
Discount factor at rate of 15.0%	*1.0000*	*0.8696*	*0.7561*	*0.6575*
Present value of marginal cash flow, $	(30,000)	13,043	11,342	9,863
Net present value, $	*4,248*			

Note: Numbers in parentheses indicate negative values.

3. Cost ranking is used for shallow surface mines in bedded deposits, e.g., coal, mineral sands, phosphates, oil shale, tar sands.

The three techniques are outlined in the following subsections. They deal primarily with the marginal costs of mining—the extra returns from the extra ore as the mine is extended laterally or deeper than with the base case plan. Capital costs and costs common to the whole operation (e.g., administration and marketing costs) are usually excluded. The impact of these costs is discussed later in this chapter.

Cutoff Grades

Cutoff grades are the key economic determinant governing the strategy for development of most underground hard-rock mines and mines where the grade of the orebody changes gradually over distance. Cutoff grades are extremely useful guidelines for short-term production operations. As grades decrease away from the main (higher-grade) ore, the cutoff grade determines whether the "next" block of material to be mined (either deeper or laterally) is classified as ore or waste (or something else).

For material to be classified as ore, then at a minimum the extra return from the extra effort (to mine it, mill it, treat it, etc.) must exceed the extra costs incurred. The *average* costs have no bearing.

Nevertheless, what constitutes an extra cost is seldom easy to define. Extra costs are also a function of the time frame over which the assessment is made.

Example 6.1:

If the mine, mill, and all downstream activities are evenly matched for production, then in the long term, changes to the cutoff grade must be sufficient to cover the extra capital costs of mining equipment and of plant establishment to process the extra material, as well as all of the extra operating costs.

Alternatively, if one section of the mine goes out of commission for, say, 3 months and the entire enterprise is suddenly limited by mine output, then the extra costs in this case are different than what they would be in the long-term case. For the duration of this period, a mining company might have to maintain all of the labor in the mine and the plant, and it might have to incur 100% of its other "operating" costs as well (e.g., demand charges for electricity). Lowering the cutoff grade for the still-active part of the mine will result in some material previously classified as waste now being classified as ore. Up to the limit of the plant capacity, this is viable provided the extra revenue after processing the reclassified material exceeds the real extra costs in the plant—i.e., exceeds the costs that are in excess of those that would have to be borne anyway.

In the preceding example, the mine may incur a loss during the period when the low-grade ore is being processed, but it will face a bigger loss if the ore is not processed and sent to waste.

Clearly, correct determination of the cutoff grade is critical at the start of the project since this determines the overall economics. The marginal costs at the start of the project include just about everything. Once a project is in operation, the cutoff grade—determined still by the marginal economics—excludes many of the now-fixed costs. If the long-term choices have been correctly made, then costs of production will be recovered by product selling price. Yet this does not imply that the mine is profitable. As demonstrated in Example 6.1, correct short-term cutoff grades mean only that the mine is *as profitable* as it can be. The optimum cutoff grade may be only the best of several money-losing alternatives.

The theory and application of cutoff grades is set out more comprehensively in Lane (1991).

Cost Ranking

Cost ranking is the key economic tool for valuing reserves and controlling the strategy for development of most shallow surface mines—particularly surface coal mines. As mines progress into deeper overburden cover or as grades (or seam thicknesses) change, the cost ranking determines whether the "next" block of material to be mined (either deeper or laterally) is (1) classified as ore or waste (or something else), (2) considered economic or not economic, and (3) included in or excluded from the mine plan.

For so-called ore (in a technical sense) to be included in the plan as economic, then at minimum the extra return from the extra effort (to mine it, process it, etc.) must exceed the extra costs incurred. The *average* costs have no bearing.

As with the cutoff-grade assessment, the time frame over which the cost-ranking assessment is made influences the decision.

Example 6.2:

If the mine, processing plant, and transportation are evenly matched for production, then in the long term, the returns from mining lower-grade (lower-yield) ore must be sufficient to cover the extra capital costs of mining equipment to handle the greater tonnage, as well as the extra capital costs of plant establishment to process the extra feed tonnage. In this case downstream capital costs can be excluded since there would be no extra tonnage of product. Of course, the operating costs of mining, processing, and transport would have to be covered.

Alternatively, if the mine, processing plant, and transportation are evenly matched for production, then in the long term, the decision to classify deeper but similar-grade material as "ore" would have to cover only the extra capital of the extra overburden-stripping equipment, plus the operating costs of the whole mining, processing, and transporting activity.

Alternatively, if one section of the mine goes out of commission because of flooding for, say, 3 months and the entire project is suddenly limited by mine output, then what constitutes an extra cost in this case takes on an entirely different complexion. For the duration of this period, the mining company might

> have to maintain all of the labor in the mine and the plant, and it might have to incur 100% of its other "operating" costs as well (e.g., demand charges for electricity). In this instance, the economics of mining higher-cost material in the still-active part of the mine will result in some material previously classified as noneconomic now being classified as economic. For example, you can use contractors for waste removal at a cost that might have previously been considered uneconomic. Up to the limit of the plant capacity, this is viable so long as the extra costs of mining plus the real extra costs in the processing plant—i.e., the costs over and above those that would have to be borne anyway—do not exceed the returns from the sale.

In the preceding example, the mine may incur a loss during the period when the contractor waste stripping is being undertaken, but it will face a bigger loss if the contractor work is not undertaken, so the economics still favor this action.

Clearly, correct determination of the limits of economically viable ore is critical at the start of the project since this determines the overall economics of the project and its overall profitability. The marginal costs at the start of the project include just about everything. Once a project is in operation, the marginal economics then exclude many now-fixed costs. Just as demonstrated in Example 6.1 and the third alternative in Example 6.2, short-term decision making and the optimum economic limits of mining may merely define the best of several money-losing alternatives.

Runge (1988) sets out a description of the application of cost-ranking techniques to an open pit coal project.

Pit Optimization

Pit optimization is the key economic tool for valuing reserves and controlling the strategy for development of most open pit hard-rock mines. As mines get deeper and as grades and other orebody characteristics change, the pit optimization technique determines whether the "next" block of material to be mined (either deeper or laterally) is (1) classified as ore or waste (or something else), (2) considered economic or not economic, and (3) included in or excluded from the mine plan.

For material to be included in the plan as economic, then at a minimum the extra return from the extra effort (to remove the extra waste above it, mine it, process it, etc.) must exceed the extra costs incurred. The *average* costs have no bearing.

Pit optimization tools have traditionally been used primarily as long-term planning aids, with shorter-term economic comparisons being done by using cutoff-grade techniques. For the long-term design, pit optimization ensures that when a company decides to stop mining in one particular direction, the *relative* economics of the ore on the boundary are the same as the relative economics at the limits of mining on any other boundary.

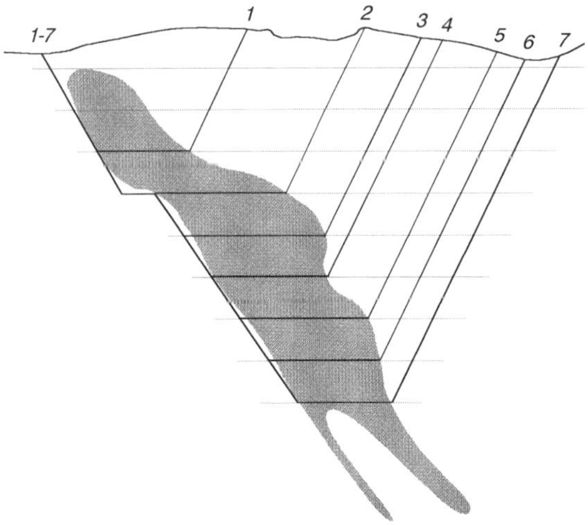

FIGURE 6.2 Blocks to be removed to uncover successive blocks

Various algorithms have been written to calculate the optimum open pit for an orebody. Of these techniques, the Lerchs-Grossman (LG) technique is the most widely accepted. The method takes a block model of an orebody and determines which blocks should be mined to obtain the maximum dollar value from the pit. This set of blocks then defines the optimum outline for the pit. The method is particularly suited to massive disseminated orebodies, such as low-grade porphyry copper deposits or deposits with multiple connected or unconnected ore zones. Simpler orebodies can be handled equally well with the technique, but they can also be understood by less-robust techniques. Two types of information are required to calculate the optimum pit.

1. For each block in the model, the user of the LG technique needs to know what blocks in a relative sense must be removed to allow the block under consideration to be mined. Usually the precedent blocks are a function of the allowable pit wall slope, but other constraints can also be applied. More than one pit slope may be allowable depending on whether the block is on the hanging or footwall or whether a ramp might need to be included. An inverted cone of precedent blocks, perhaps subject to minimum pit width constraints, is defined as shown in Figure 6.2.
2. The economic factors will then define a block as being either positive (a revenue generator) or negative. Economic factors that are considered include mining, haulage, and processing costs as well as ore revenue and recovery factors.

Once the slope and cost parameters have been applied, the LG algorithm is applied, producing an optimum pit for the given constraints.

To reflect more closely the time-value-of-money factors, some programs use a sensitivity analysis to generate a family of optimal pits given a range of costs and revenues. From the optimum pit of each scenario, a mining schedule of cash flows may be produced. Multiple ore-processing methods and multiple ore types can be handled.

The original development of the LG method was undertaken by Lerchs and Grossmann (1965). Subsequent work by J. Whittle and others (see, for example, Alford and Whittle [1986], Whittle [1988], Whittle and Rozman [1991], and Roditis [1993]) have extended the theory and application of this technique to account for the time value of money in sequenced excavation.

CAPITAL VALUES AND DEVELOPMENT STRATEGIES

Each of the three techniques—cutoff grades, cost ranking, and pit optimization—leads to mine plans that maximize the value of the resource within certain constraints. However, there is nothing to ensure that this maximum value is actually *positive.* The mine does not need to be profitable for one to undertake these sorts of analyses on it.

This leaves at least three important issues unaddressed:

1. The origin of costs. The techniques are largely concerned with operating costs. The extra costs of mining extra ore are usually the operating costs. The initial application of the techniques requires the user to *assume* certain operating costs. However, the operating costs are a function of the equipment and methods deployed, and the equipment and methods deployed are a function of production, and production is a function of reserves. In practice, the equipment suited to many deposits *can* be quite clearly defined in advance, yet this still doesn't solve all of the problems. Truck haulage by contractor implies the whole cost is an operating cost (the rate includes an amount that the contractor applies to repaying his or her capital), but truck haulage using identical *owned* equipment has a lower operating cost because it excludes capital repayment components.
2. Accounting for fixed costs. The techniques include as "ore" everything that makes a positive contribution to the value of the mine—but they are based on an implicit assumption that the mine is already established. If the extra revenues received (from extending the limits of mining) exceed the extra costs, the block should be included. Yet there is no indication whether the surplus is sufficient to cover the fixed costs or even the increase in fixed costs (see Example 6.3 after this list). Also, there is no indication whether the possible higher returns from alternative schemes might justify even higher fixed costs.
3. Extraction sequence. The techniques map out economically viable ore, but they normally do not indicate the optimum sequence for extracting this ore. A mine plan focused on higher-grade ores early in the mine life may yield a higher net present value, *even if this means wasting viable but lower-grade ores.*

Example 6.3:

Access to 10 additional strips might require additional ramp development, the cost of which cannot be borne by any one strip alone. Allocating one-tenth of the development to each of the strips is inappropriate (except as an *ex post* accounting measure) because the ramp has to be in before the first strip is mined. Once the ramp has been developed, other blocks will be viable to mine even if they are unable to shoulder their proportionate cost of ramp development.

The issues mentioned in the preceding list are addressed in the following subsections.

Limiting the Mining Options Available

In a mine plan with thousands of potential mining blocks, assessment of the capital and operating cost trade-offs and optimum development strategy is not a trivial process. For practical purposes, it is usually addressed in two steps:

1. Establish the maximum and minimum production rate and capital cost that would apply to the mine given its known or assumed cost structure. A mine that requires the construction of a new railway and port is unlikely to be viable at low production rates, for example.
2. Within the limitations just mentioned, establish the key constraints having a significant influence on the mine's development and economics. This usually reduces the possible choices much further. For example, large, electrically powered equipment may be infeasible in remote mine sites with no connection to the power grid.

Each potential mining method has different maxima and minima for both production and capital investment. The deposit will have to be considered at least as many times are there are different mining methods to exploit it.

Example 6.4:

A potential mine for which the economics have been assessed assuming dragline operating costs would have greater reserves and potentially higher rates of production than the same mine assessed based on shovel/truck operating costs. This doesn't mean that the dragline case is necessarily more profitable. Because the operating costs per volume of waste moved by dragline are generally less than shovel/truck waste operating costs, the reserves will be greater. Whether the mine is more economical will depend on whether the additional surplus (price less variable cost) is sufficient to cover the additional capital in the dragline method.

Even for fixed mining techniques, different production rates may imply different *sizes* of equipment, with significant changes in cost.

Example 6.5:

A dragline technique will have a minimum size—equivalent to just one dragline. Anything smaller than this will require a smaller dragline, for which the operating costs and technical limitations may be quite different.

Most mines, if planned with higher production rates, also exhibit lower costs. But the cost/production rate function is seldom a smooth curve. There may be *increments* of production that are so important in economic terms that the mine should logically be developed around these increments.

Example 6.6:

Suppose an existing railway has a maximum capacity of 500,000 tpy. Beyond this, a new railway would have to be built. Because any new railway has high fixed costs, the "next step" would demand production of at least 5 million tpy to be a viable option.

In practice, the range of options available usually reduces to a manageable number of iterations. The number of iterations can be further reduced by the scrutiny of capital requirements, a description of which is set out in the next subsection of this chapter.

Most large-scale mining is capital-intensive, and full utilization of high-capital-cost equipment is important. This does not mean that if capital cannot be fully utilized the mine is likely to be uneconomic. An assumption that increments in capital strongly dictate mine development strategies may be an error. Chapter 10 sets out an example of capital utilization demonstrating how an underutilized large machine can still be more economical than a fully utilized smaller machine because of the lower operating costs per unit of material moved.

Example 6.7:

A dragline strip mine is not necessarily bound to production from integral numbers of draglines. For instance, production equivalent to 2.5 (larger) draglines may have similar economics to production from an exact 3 (smaller) draglines, even though the third (larger) dragline is only 50% utilized.

Amount of Capital a Mine Will Support

At such an early stage of assessment of a mining project, capital investment requirements are difficult to estimate. Nevertheless, once the operating costs of each block of ore and waste are known, then the maximum amount of capital that the mine will support can be determined.

Most of the capital is expended at the start of the mine, and the highest return on this capital is achieved by a mine plan for which the greatest cash flows occur early in the mine life. Assuming it is feasible, the greatest return on investment will be realized if the lowest-cost mining blocks are mined first, progressing to higher-cost blocks later in the mine life. Ideally, the block to be mined last would be the one for which the operating costs were just below the selling price, because this is the block making almost no contribution to paying off the mine capital.

The maximum amount of capital a mine will support is determined by the net present value of the surplus cash flows when reserves are exploited in sequence from the lowest cost to highest cost.

Figure 6.3 shows an example plot of cumulative mining reserves arranged in increasing order of mining cost (for a certain assumed mining method). The mining costs in this case are the total direct operating costs of the mine

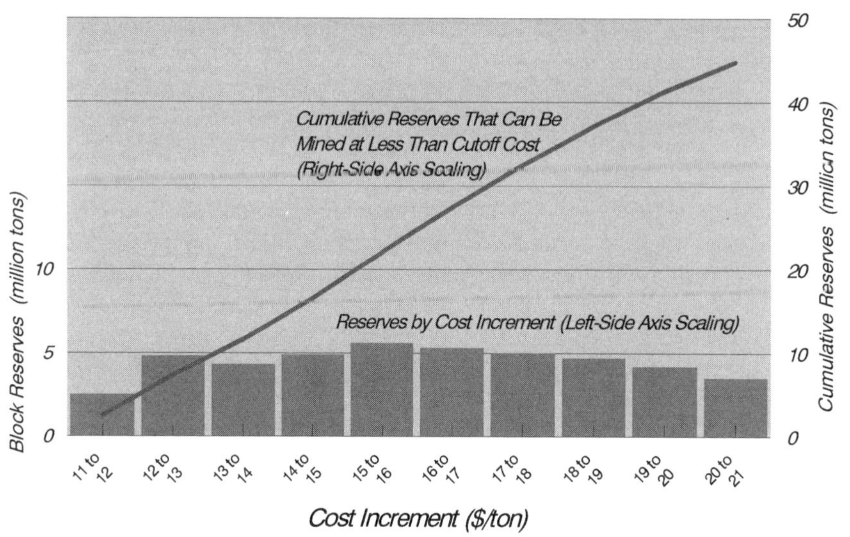

FIGURE 6.3 Example plot of cumulative reserves less than cutoff mine-operating cost

TABLE 6.4 Production schedule for mining according to the cost/tonnage curve of Figure 6.3

	Year					
	1	2	3	4	5	6
Production, *million tpy*	2.4	4.0	4.0	4.0	4.0	4.0
Average direct costs, *$/t*	11.50	12.48	13.27	14.20	14.97	15.57
Overhead operating costs, *$/t*	2.08	1.75	1.75	1.75	1.75	1.75
Total operating costs, *$/t*	13.58	14.23	15.02	15.95	16.72	17.32

assigned to the blocks, excluding capital, administration overhead, and marketing costs.

In some mines it is not possible to sequence the lowest-cost ore ahead of the higher-cost ore—if the reserves occur together, mining the lowest-cost ore might destroy access to the higher-cost ore. Yet a surprisingly large number of mines can be developed this way if so desired, e.g., simple (single-seam) open pit coal mines, phosphate mines, beach sand mines, most bauxite mines, and other alluvial deposits where the mine is relatively shallow and has a large lateral extent.

Regardless of whether the lowest-cost-ore-first sequence is possible or not, there is no doubt that the present value of the cash flow—and hence the amount of capital investment the project will support—is maximized in this scheduling sequence. Scheduling the grade/tonnage (or cost/tonnage) curve places an upper limit on the capital.

The first 6 years of production from the example deposit used for Figure 6.3 are shown in Table 6.4, based on an annual production rate of 4 million tpy

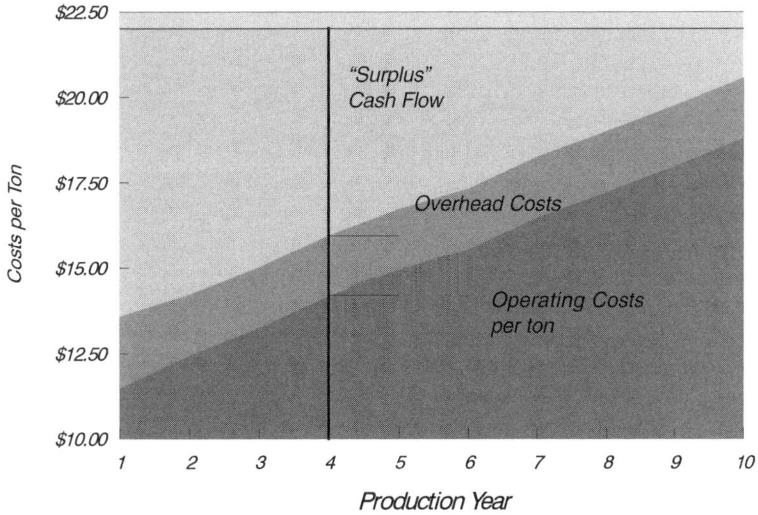

FIGURE 6.4 Cost of mining by year from lowest-cost to highest-cost mining

TABLE 6.5 Present value of surplus cash flow: 4 million tpy production, discounted

	Year					
	1	2	3	4	5	6
Production, *million tpy*	2.4	4.0	4.0	4.0	4.0	4.0
Total operating costs per unit, *$/t*	13.58	14.23	15.02	15.95	16.72	17.32
Total operating costs, *million $*	32.60	56.90	60.10	63.80	66.90	69.30
Total revenues at $22/t, *million $*	52.80	88.00	88.00	88.00	88.00	88.00
Operating surplus to apply to capital, *million $*	20.20	31.10	27.90	24.20	21.10	18.70
Discount factor at 15% return on investment	*0.8696*	*0.7561*	*0.6575*	*0.5718*	*0.4972*	*0.4323*
Present value of "surplus" cash flow, *million $*	17.57	23.52	18.34	13.84	10.49	8.08
Present value of 10 years of production = $105.37 million						

(except only 60% of this production in year 1). The direct operating costs in Table 6.4 are the costs that were assigned to the blocks in the cost-ranking/cutoff analysis. Table 6.4 also shows the addition of overhead operating costs—in this case an amount of $2 million annually, plus $1.25/t. The total operating costs are shown and plotted in Figure 6.4.

The difference between the selling price and the cost of production is the surplus that is available to pay back the capital. Table 6.5 shows the first 6 years of production again, calculating the surplus cash flow available with the selling price of $22/t and expressing the whole 10-year cash flow in present value terms. At a production rate of 4.0 million tpy, the project is capable of supporting a maximum initial capital investment of $105 million—assuming that an optimum mining sequence is possible.

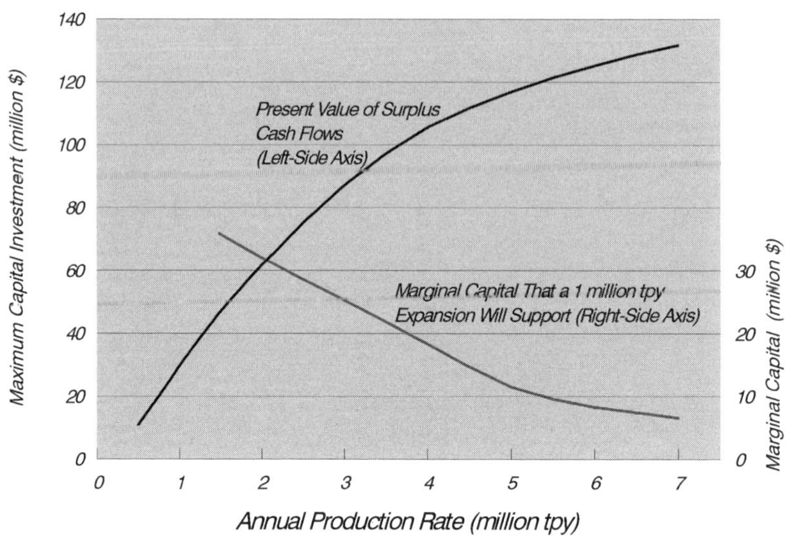

FIGURE 6.5 Maximum capital investment for a project

At lower production rates, the project has smaller cash flows and will therefore support less capital—however, less capital is needed at these lower rates. Figure 6.5 shows the same project analyzed at production rates varying from 500,000 tpy to 7 million tpy, showing the maximum capital investment possible for these rates of production.

The calculation in Table 6.5 shows that a production rate of 4 million tpy will support a maximum of $105 million of capital. Since this is the best possible sequence for maximizing the present value of the cash flow, if *at this stage* a mine cannot be developed for less than that amount of capital, then there is little point in proceeding further. From a strategic viewpoint, there are two options available:

1. Examine alternative production scenarios assuming the same basic equipment.
2. Reexamine the whole mine on the basis of some alternative mining scheme. This will require reevaluation of the costs, cutoff, reserves, and potential mining schedules.

A clue as to the optimum production rate can be derived from the marginal capital requirement. The right-side axis of Figure 6.5 shows this marginal capital calculation—the extra capital that is supported for each 1 million tpy increase in annual production.

In the example, at the base case production rate of 4 million tpy, the incremental capital to expand or contract production by 1 million tpy is $18.3 million. If the incremental capital supported by the higher tonnage is more than the capital needed to produce the additional 1 million tpy, then the economics of the mine would be improved by adopting a higher production rate. Conversely,

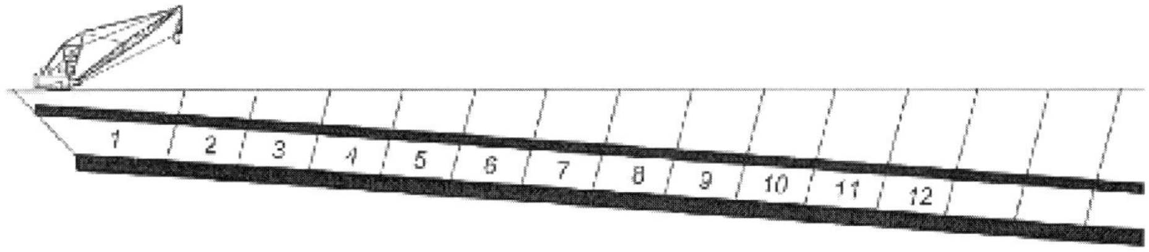

FIGURE 6.6 Cross section of mine with increasing overburden depth

if the incremental capital supported by the higher production is less than the extra cost of equipping the mine for that higher production, then the economics of the mine would be improved by adopting a lower production rate.

Note that the effect of taxation has not been brought into the preceding assessment. At this early stage of evaluation, taxation can usually be ignored, since the objective is simply to determine whether the project is likely to be viable or not. The imprecision of the base data means that a *clear* result should be achieved. If the project shows limited attractiveness at this early stage of evaluation, then caution is the order of the day. If the project is borderline, then in all probability the profit will be close to zero—and hence tax will be zero. Therefore, the decision *not* to include tax in this analysis is correct for the purposes of the decision-making guidance sought from the evaluation.

Impact of Production Rate

The discussion of initial assessments earlier in this chapter focused on particular parts of the mine, examining them on a stand-alone basis to decide whether they are viable to mine or not. This section focuses on the entire mine—and examines the influence of mining sequences on the economic ore definition. On a stand-alone basis, a block may be profitable, but there may be a mutually exclusive and *even better* mining sequence that precludes mining of the block. The decision to include it or exclude it from the reserve estimate is also a function of the mining schedule.

The effect of scheduling on the economics of mining is best illustrated by an example. The example pertains to a simple two-seam dragline mine starting at a depth of only 5 m of overburden and extending to 60 m of overburden, as shown in Figure 6.6. There is a constant 25 m of interburden between the seams. The numbers in the figure represent phases of excavation; phase 1 must be mined (or abandoned) before any mining can commence in phase 2, and so on. The dilemma is visually obvious from Figure 6.6. The first few blocks in the lower seam are quite economical to mine. However, if they were ignored, the early cash flow from mining focused *only* on the upper seam (initially) may be even more profitable.

Table 6.6 shows the calculated costs associated with mining this deposit with dragline, reflecting the costs of mining the upper seam *as if it were the only seam mined.* The lower-seam costs are genuine marginal costs—change in total

TABLE 6.6 Cost of production: two-seam mining case

	Phase											
	1	2	3	4	5	6	7	8	9	10	11	12
Overburden thickness, *m*	5	10	15	20	25	30	35	40	45	50	55	60
Upper coal thickness, *m*	4	4	4	4	4	4	4	4	4	4	4	4
Interburden thickness, *m*	25	25	25	25	25	25	25	25	25	25	25	25
Lower coal thickness, *m*	6.7	6.7	6.7	6.7	6.7	6.7	6.7	6.7	6.7	6.7	6.7	6.7
Upper coal, product, *million tons*	5.13	5.13	5.13	5.13	5.13	5.13	5.13	5.13	5.13	5.13	5.13	5.13
Lower coal, product, *million tons*	8.59	8.59	8.59	8.59	8.59	8.59	8.59	8.59	8.59	8.59	8.59	8.59
Total, *million tons*	13.72	13.72	13.72	13.72	13.72	13.72	13.72	13.72	13.72	13.72	13.72	13.72
Operating costs, mining to upper seam, *$/t*	2.75	3.50	4.25	5.00	5.74	6.49	7.73	8.70	9.77	10.89	12.14	13.45
Operating costs, mining of both seams, *$/t*	3.80	4.29	4.73	5.06	5.52	6.02	6.54	6.96	7.42	7.85	8.29	8.73
Extra costs of mining to the extra seam, *$/t*	4.43	4.77	5.02	5.13	5.39	5.73	5.83	5.95	6.01	6.04	6.04	6.04

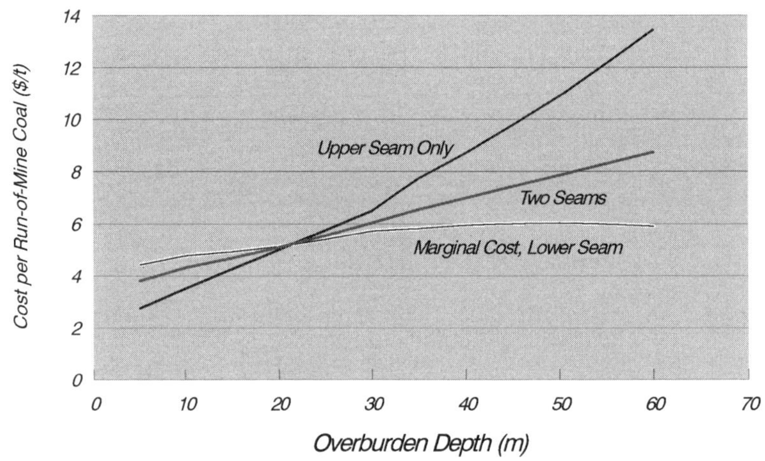

FIGURE 6.7 Costs of mining for the two-seam dragline mine

costs. They have been estimated by subtracting the total single-seam costs from the total two-seam costs and applying the cost increment to the tonnage in the lower seam.

In Table 6.6 each phase does not necessarily represent a mining strip. Data are presented in 5-m-depth increments, which correspond to several mining strips. Figure 6.7 shows the information from Table 6.6 in a graphical form. In the figure, as the mine advances downdip into deeper overburden cover, the cheapest coal at that particular part of the mine is indicated. If one overlooks for a moment the fact that mining has to be undertaken in discrete-sized blocks or strips, then until the overburden thickness reaches about 20 m it is most economical to mine *only* down to the upper seam—mining to the lower seam results in less profit. When the overburden is more than 20 m thick, the economics of the mine are *enhanced* by mining to the lower seam. If the cutoff were less than \$4.43/t, then the decision would be easy—only mine the upper seam, and stop mining altogether after phase 3. If the cutoff were, say, \$8.00/t, the scheduling dilemma becomes evident. Mining the lower coal in the first few strips will be profitable, but *not* mining them will be *even more profitable,* at least initially (but will result in lower profits later in the mine life).

Consider, for example, the case from Figure 6.7, where the operating costs associated with mining at an overburden depth of 15 m and 30 m are as follows:

		Operating costs (\$/ton)	
	Quantity (t)	Overburden depth = 15 m	Overburden depth = 30 m
Upper seam (quantity and average cost)	5,130,000	4.25	6.49
Both seams (quantity and average cost)	13,720,000	4.73	6.02
Lower seam (quantity and marginal cost)	8,590,000	5.02	5.73

The lower-seam marginal per-ton cost derives from subtracting the total cost of the upper seam from the total cost of both seams, then dividing by the quantity of coal in the lower seam. At 30 m of overburden thickness, if it is viable to mine at all, mining *should* go down to the lower seam. Regardless of the cutoff cost, the lower seam makes mining more profitable. At 15 m of overburden thickness, a strategy to maximize profits suggests that mining *should not* go down to the lower seam. Yet if a mine operator followed this strategy, an apparently inconsistent result follows. Early in the mine life, coal costing \$5.02/t is being passed over, whereas when the mine gets deeper, coal costing \$5.73/t is being included.

In practice, the decision to mine the higher-cost coal earlier in the mine life is made based on when, later in the mine life, the company might be forced to mine *even-higher-cost* coal. It is a decision based on time as well as direct cost.

To resolve this type of problem, two scheduling sequences for production from the mine are prepared. The scheduling sequences show when the blocks will be mined, so as to allow comparison of the one-seam-initially and two-seams-always alternatives. At low production rates, advance through the deposit is slow. A lot of time will elapse before higher-cost coal is encountered. Figure 6.8 shows a characteristic low-production scenario following both the one-seam-initially case and the two-seams-always case.

In Figure 6.8, two scheduling sequences have been prepared:

- For the first scheduling sequence, labeled "always mining the cheapest coal," only the upper seam is mined until a depth is reached at which the extra cost of mining to the lower seam is more economical, and thereafter both seams are mined.
- For the second sequence, labeled "mining all coal," both seams are mined right from the start.

The sequence that results in greatest profit is the sequence with the lowest cost of production when expressed in *present value* terms. A discounted average

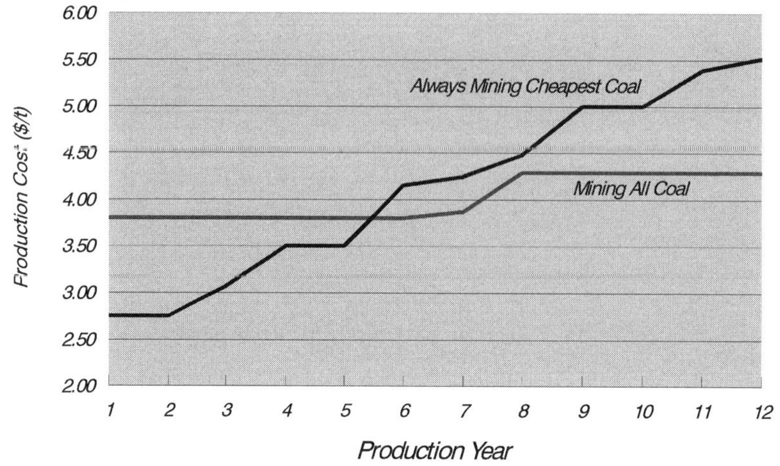

FIGURE 6.8 Mining costs over time: low-production case

TABLE 6.7 Scheduled production costs by year: two cases

	Low-Production Case: 2.0 million tpy											
	Year											
	1	2	3	4	5	6	7	8	9	10	11	12
Always mine cheapest coal, *$/t*	2.75	2.75	3.07	3.50	3.50	4.15	4.25	4.48	5.00	5.00	5.39	5.52
Mine all of the coal, *$/t*	3.80	3.80	3.80	3.80	3.80	3.80	3.87	4.29	4.29	4.29	4.29	4.29
	High-Production Case: 4.0 million tpy											
	Year											
	1	2	3	4	5	6	7	8	9	10	11	12
Always mine cheapest coal, *$/t*	2.75	3.29	3.82	4.36	5.00	5.45	5.52	5.52	5.74	6.02	6.02	6.02
Mine all of the coal, *$/t*	3.80	3.80	3.80	4.08	4.29	4.29	4.34	4.87	4.87	4.87	4.96	5.06

cost of production can be calculated for each of the sequences in a manner similar to the calculation in Chapter 5. The second scheduling sequence has a discounted average cost of production of $3.92/t, compared to $3.68/t for the first sequence. The *extra* costs of mining the lower-seam coal early are not warranted by the savings in mining cost later in the mine life—the overall costs are 6% higher. This is intuitively obvious from Figure 6.8 by comparing the areas under the two curves: The savings from year 5 onward are not sufficient to offset the higher costs leading up to year 5.

Table 6.7 sets out the results of 12 years of production based on production rates of 2.0 million tpy (the low-production case shown in Figure 6.8) and 4.0 million tpy (the high-production case).

Figure 6.9 shows information similar to the situation in Figure 6.8. However, Figure 6.9 plots the costs of production for the mine scheduled at the high production rate (4.0 million tpy). In Figure 6.9, if the lower-seam coal is *not* mined at first, the mine is forced into *even-higher-cost* coal sooner. The

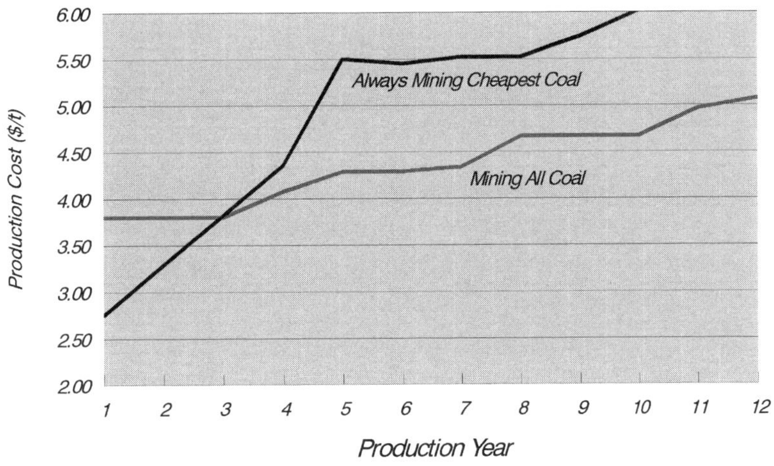

FIGURE 6.9 Mining costs over time: high-production case

discounted average cost of production for the "mining all coal" sequence ($4.179/t) is actually 6% *lower* than for the alternative sequence ($4.454/t)—even though for the first 3 years, coal-mining costs are higher than they otherwise could have been. The first 3 years of higher-cost mining represent a form of capital (voluntarily incurred higher operating costs) that is repaid through lower operating costs later.

The preceding example has deliberately been kept simple to illustrate the time-based trade-offs; however, in actual application, many of the comparisons become even more critical. Scheduling the progressive mining out and backfilling of underground stopes (when there is still profitable ore remaining) is another common application of the same economic problem. A third example common in all kinds of mining involves blending. The true value of a block is determined not only by its own characteristics, but also by the characteristics of the blocks with which the block under consideration can be contemporaneously mined.

Whether a particular seam, whole section of a mine, or stope extension into lower-grade ore becomes viable can be dependent on the scheduled production sequence and rate. The inclusion or exclusion of these reserve blocks can have a large impact on the classification of viable reserve quantities.

This time-based inclusion or exclusion from the economic ore reserve highlights a characteristic not commonly recognized. Compared to low-production mines, mines operating at high production rates may have more extensive reserves than other mines, or even the same mine, *despite identical operating costs.*

SUMMARY

The economic assessment of mineral deposits does not apply only to operating mines or detailed feasibility studies. It applies to undeveloped reserves and is a critical factor in establishing the value from exploration effort.

Exploration effort and the value of reserves in the ground are very sensitive to the economics of mining. If mining costs are uncertain and represent a significant proportion of the selling price, then large reserves or reserve extensions have little value unless they also facilitate lower mining costs. The focus for exploration must be on refinement of reserves and proving up of technologies. If technologies are well understood and there is opportunity for additional reserves to be brought into production quickly, then the value of reserves (or potential reserves from additional exploration effort) may easily justify high costs. Production rates, the *variability* in production rates, and timing of cash flows become critical.

Once potentially viable deposits are identified, initial assessments are aimed at establishing orebody shape from an economic perspective—what is included in the reserve or not. These initial assessments are primarily based on operating costs. A cost-ranking analysis or pit optimization study is not a mine plan or mine design. It is an understanding of the economics of the project *assuming* a mine plan is viable. It does not provide definitive support for risk-based decision making where the mine plan has some uncertainty with respect to its implementability.

A reserve block is viable in any initial assessment if the *extra* economic returns from mining it exceed the *extra* economic returns from the best plan that *does not* include it. The "extra" costs or returns are often difficult to define and (particularly in operating mines) frequently change on a day-to-day basis. "Optimum" mine plans are not necessarily profitable.

Capital cost estimates for mineral deposits are available only after comprehensive study, but the maximum allowable capital (for any mine that is to be economic) can often be estimated early in the evaluation process. Early estimates derived from the present value from surplus cash flows provide a valuable guideline for strategic planning and for refinement of exploration targets where substantial differences in capital apply to alternative possible mining schemes.

The economics of reserves in the ground are also a function of the production rate. Even for the same costs of production, faster rates of production imply increased reserves because this approach decreases the incentives to pass over low-profit reserves. Higher production rates also commonly mean lower costs of mining. New mines being developed with a view to expansion are susceptible to incorrect reserve definition—early reserves that appear unattractive initially (when low production rates underpin planning priorities) may become quite attractive *even before the mine commences* if planning priorities change toward higher rates of production.

CHAPTER 7 Ownership Costs and Capital Costs

Every economic evaluation involves use of mining equipment—whether fixed or mobile. There is a capital cost of owning this equipment, as well as operating costs of running it. This chapter looks at the ownership cost.

Sometimes the ownership cost calculation is no harder than obtaining a quotation from the supplier and perhaps coupling it with a quotation from a leasing company. Alternatively, processing plants and civil infrastructure works often include components derived from many different sources, and costs must be built up from first principles and adjusted for currency effects. Often it is not the initial capital cost that is required but rather an *annualized* equivalent cost. For example, if one option has equipment that lasts just 3 years, how can this option be compared with alternative equipment that lasts 5 years even though it is needed for only 3 years?

This chapter sets out how to understand the equivalent costs of *owning* equipment. It involves the capital cost, life of the machine, and other costs of ownership. Three alternative methods are presented for working out the hourly ownership cost. The average investment method is the first method this chapter describes; it is the simplest but least precise calculation. The equivalent lease cost method is also described; it is a method that yields practical and reliable results for single items of equipment and for cases where taxation does not need to be considered. The most reliable method, the discounted average cost method, is described last and may be used for any investments and tax treatments.

MACHINE LIFE AND CAPITAL COST

Before the three methods are described in detail, an important distinction must be made among the terms *accounting life, operating life,* and *economic life.*

Accounting life is the life over which the machine is depreciated for tax purposes. It is the rule that the tax authorities permit a company to use in working out its taxes. Some companies adopt more than one rule when determining the accounting life, particularly if they believe that the rule for tax calculations is inappropriate for their own application. In this case, the (second) accounting life is aimed at valuing the company's assets in a manner consistent with the market value of the asset in an ongoing operation. Decisions to buy and continue using equipment are based on many more factors than encapsulated in the simple accounting life calculations; hence, the accounting life may not represent the expected value of the machine throughout its life.

The operating life is the life over which the mine operator expects the equipment to operate at an availability satisfactory to meet production targets. The operating life is usually similar to the economic life because, if manufacturers are finding that their machines are being superseded before those machines are worn out, then over time they change the design to bring the operating life into line. Nevertheless, with fast-paced new machinery development, there can often be considerable divergence.*

The economic life of equipment is the life over which it can undertake its defined task more cost-effectively than alternative ways of accomplishing that task. In a world of unpredictable change, equipment that is less flexible or less adaptable is more susceptible to shortened economic life than is more flexible equipment. Thus, bucket-wheel excavators and conveyor-based mining schemes (fairly inflexible machines) find application primarily in mines operating under long-term contract to domestic customers—mines that do not require a lot of adaptability. On the other hand, mines operating in the export markets that are much more volatile and changeable commonly favor flexible mining equipment, such as trucks and shovels. Table 7.1 sets out the typical operating life of common mining equipment.

For an estimate of the (intended) life of mining equipment, the shift schedule should also be considered. A grader that works only on day shift, for 1,250 operating hours per year, has less than 20,000 operating hours after 15 years and will commonly be retired after this time. Its effective operating life is a function of time. The same grader working multiple shifts could be quite productive for over 30,000 hours accumulated over 8 to 10 years. Purchasing data should also be tabulated and includes

- the origin of the quote
- the list price free on board (FOB) factory
- the country of origin of the equipment and the exchange rate
- transportation and insurance charges
- the delivery time
- the payment schedule
- erection costs subdivided into labor and materials
- the spare parts (or spares) holding

* Personal computers are an example of this phenomenon. Older machines are as functional *at performing the tasks for which they were designed* as when they were first installed, but newer machines can do this much and more at lower cost. The effective *economic* life of the computer is a function not of its availability or any other physical attribute of the computer, but rather of the availability of *even better* substitutes.

TABLE 7.1 Typical operating life of mining equipment

Item	Number of Hours: Poor Conditions	Number of Hours: Average Conditions	Number of Hours: Good Conditions
Dozer	18,000	25,000	35,000
Grader	20,000	30,000	50,000
Large front-end loader	20,000	30,000	45,000
Hydraulic excavator (small)	15,000	22,500	30,000
Hydraulic excavator (large)	20,000	30,000	45,000
Scraper	12,000	16,000	20,000
Truck (50 to 100 t)	20,000	30,000	42,000
Truck (large)	30,000	45,000	60,000
Rope shovel	60,000	80,000	100,000
Walking dragline	60,000	100,000	150,000

Figure 7.1 presents a sample layout of the buildup of capital costs for a hydraulic excavator. The layout recognizes the requirement for working capital for spares associated with individual items of equipment. This cost is not normally assigned to the equipment; rather, it is separately tabulated in whole-project cash flows. The amount of spares holding is a function of

- How critical the machine is. If the machine is allocated to product load-out and is the only machine on-site, then it is vital for the machine to be available when needed. If the machine is assigned to reclamation, the primary concern is with how much work it does over, say, 1 year rather than whether it is available on a particular day.
- The length of time it takes to get spares. A well-stocked adjacent supplier facility may allow spares holding to be reduced substantially. For some spares (e.g., large dragline gears), the delivery time may also include manufacturing time.
- Consignment stock. The spares holding refers to the mining company's own investment in spares. Suppliers will sometimes place spare parts on consignment in the mining company's warehouse; in such cases, the mining company pays for the parts only when it uses them.
- Spares-sharing arrangements. If there are a lot of mines in the same area and there is a lot of equipment similarity, then an agreement may be made for spares sharing. This sort of sharing would not be applicable to spares used on a regular basis, such as dozer grouser plates and ripper boots, but it may be important for "insurance" spares such as dragline sheaves and walking gears that may not even be in stock in the factory.

Figure 7.1 sets out the calculation of the initial cost of the equipment and is a relatively straightforward tabulation. Often it is necessary to present costs not just as an initial amount of capital but as an equivalent hourly "capital" or ownership cost. This sort of calculation is more complicated. The three methods described in the remainder of this chapter may all be used. The simplest methods ignore time value and tax considerations and are therefore somewhat imprecise, but they are also easier to calculate. Whether the

Machine and Model: Hydraulic Excavator, Model AAA

Estimated Life Characteristics:
Estimated Life: .... 10 years
Annual Usage: .... 3,000 hours
Life for Costing: .... 30,000 hours
Salvage Value: .... Nil

Quotation: Obtained From the Local Equipment Supplier

Origin of Equipment: Built in Europe

Purchase Price:

Item	Foreign Currency (€)	Local Currency ($)
Price FOB	2,310,750	—
Extras	—	395,000
Freight	355,500	—
Erection		395,000
Other		197,500
Total	2,666,250	$987,500
Exchange Rate (€ per $)	0.9	
Equivalent Local Cost (of Foreign Components)	→	$2,962,500
TOTAL		**$3,950,000**

Delivery Time: Quoted 8 months

Payment Schedule:
20% at Order
40% on Shipment of Major Components
30% on Delivery
10% Retained Until Full Production

Spares Holding: 5% of Purchase Price

FIGURE 7.1 Sample layout of the buildup of capital costs

improvement in precision in the calculation is worth the effort depends on the desired level of sophistication of the study.

AVERAGE INVESTMENT METHOD

The average investment method is a method many equipment suppliers use for calculation of the annual or hourly ownership cost. This method has been included in the text only for completeness and to allow for comparison with the method used in common handbooks supplied by major equipment manufacturers. It has the advantage that it does not need a calculator to derive, but its disadvantage is that it is fairly imprecise. With the widespread use of calculators and spreadsheets, the computational ease of this method is no longer any significant advantage over the other two methods in this chapter. For simple calculations the equivalent lease cost method is recommended.

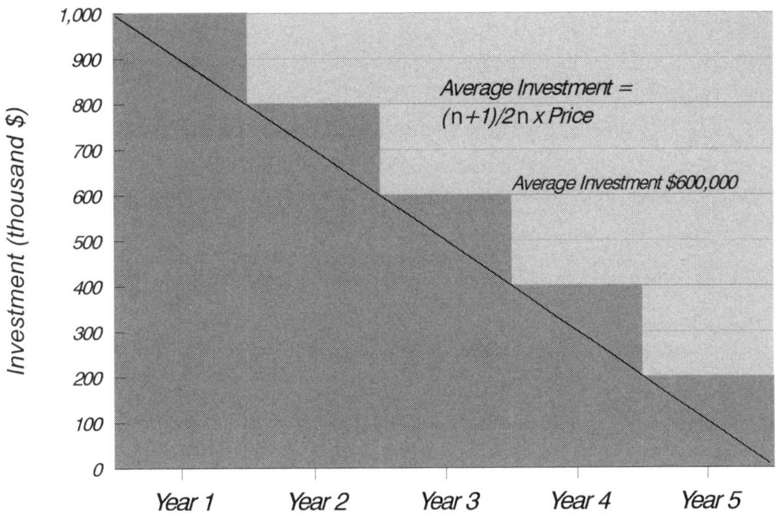

FIGURE 7.2 Average investment over machine life

The average investment method has two components: one representing the wearing out of the equipment and another representing the return on the money the company has tied up in the equipment while it is operating. The method assumes straight line depreciation and a percentage of the annual average investment to cover interest, taxes, and insurance. The basic form is expressed as follows:

$$\text{depreciation cost} = \frac{\text{capital cost}}{\text{life (hours)}}$$

$$\text{interest} + \text{taxes} + \text{insurance} = \frac{\text{average yearly investment} \times \text{rate\%}}{\text{annual operating hours}}$$

where

average yearly investment = delivered price × $[(n + 1)/2n)]$

n = the life in years

rate% = the "interest" on the invested funds in the equipment, plus an insurance amount (typically 1–2%), plus any property taxes levied on the valuation of the equipment

This calculation assumes a value of zero at the end of the machine life. If the machine is worth nothing at the end, then this suggests that the average investment should be just one-half of the delivered price. However, this method assumes annual valuations, and since equipment must be purchased *before* any production is achieved, there is an implicit step function in the size of the investment from year to year. This step function is the origin of the $[(n + 1)/2n]$ factor. Figure 7.2 shows this valuation graphically. As the figure illustrates, the $[(n + 1)/2n]$ factor takes account of the fact that the capital cost occurs ahead of any ownership costs, although it does not strictly account for any other time-value considerations. Example 7.1 tabulates how this calculation is undertaken.

Example 7.1:

A dozer that operates for 4,000 hours per year and has an expected life of 20,000 hours has an initial capital cost of $1 million. What is the hourly ownership cost, assuming the company seeks a 15% return on investment?

Initial investment (delivered price)	$1,000,000
Machine life	5 years
Average yearly investment	$600,000
Taxation	Disregarded in this method
Required return on investment	15%
Ownership cost associated with return on investment	$600,000 × 0.15/4,000 = $22.50/hour
Ownership cost associated with depreciation	$1,000,000/20,000 = $50.00/hour
Total ownership cost	*$22.50 + $50.00 = $72.50/operating hour*

For the average investment method to be used, the machine has to have a finishing value of nil. However, this does not mean that the method cannot be used to calculate ownership costs for equipment that is to be used only for part of its life. To account for this circumstance, two calculations are necessary:

1. Calculate the hourly ownership cost assuming the machine will be fully worn out over its full useful life.
2. Envisage the machine at the part-life position. What would it be worth then? If someone purchased it then and used it for the balance of its life, what would be its ownership cost for this purpose?

The ownership cost for the first half of the machine life is equal to the cost for the whole life less the cost for the second half of its life. Example 7.2 shows this calculation.

Example 7.2:

For the dozer in Example 7.1, what would be the ownership cost if it is to be used for only 3 years and then sold at its written-down value?

Answer:

The ownership cost of the dozer for the balance of its life is given by the following:

Initial investment (of the 3-year-old dozer)	$400,000
Remaining machine life	2 years, or 8,000 hours
Average yearly investment	$300,000
Ownership cost associated with return on investment	$300,000 × 0.15/4,000 = $11.25/hour
Ownership cost associated with depreciation	$400,000/8,000 = $50.00/hour
Ownership cost, remaining machine life	*$11.25 + $50.00 = $61.25/operating hour*

For the ownership costs to balance over the whole life of the machine, the ownership costs over the first 3 years must be higher than average.

Whole life	\$72.50/hour	for 20,000 hours =	\$1,450,000
Final 2 years	\$61.25/hour	for 8,000 hours =	\$490,000
First 3 years	***\$80.00/hour***	***for 12,000 hours =***	***\$960,000***

An alternative way of arriving at the same answer is to envisage the machine as "two" machines—one that costs \$600,000 and wears out completely in 3 years, and one that costs \$400,000 and never wears out over the same 3 years. The "second" machine has no depreciation, just "interest" charges on the \$400,000.

Although the average investment method is somewhat limited, for quick estimates it yields surprisingly close results with less effort than the two more sophisticated methods following.

EQUIVALENT LEASE COST

An alternative and preferred method is to treat the plant as if it were being leased. Lease rates include all of the factors considered in the average investment method, but they more correctly account for the higher interest component of the cost earlier in the equipment life. For this calculation, the primary inputs are

- interest rate, i
- present value, PV
- term in years, n

The annual payment can be calculated by using the capital recovery formula described in Chapter 5 (see Equation 5.3, p. 48). Minor adjustments may need to be made for stamp duty or other charges, but for the simplicity of this calculation they can often be ignored. This method is simpler and more realistic than the average annual investment method.

Example 7.3:

Using the equivalent lease cost method (capital recovery formula), what is the average hourly ownership cost for the same case as given in Example 7.1?

Initial investment (delivered price)	\$1,000,000
Machine life	5 years
Required return on investment	15%
Taxation	Disregarded in this method
Capital recovery factor (see Appendix A or Equation 5.3)	0.2983
Equivalent annual payment	\$1,000,000 × 0.2983 = \$298,300/year
Ownership cost (4,000 operating hours/year)	\$298,300/4,000
Total ownership cost	*\$74.58/operating hour*

As with the average investment method, the machine has a finishing value of nil. This also makes it harder to calculate ownership costs for equipment that is only to be used for part of its life. Again, the easiest way to account for this is to envisage the machine as "two" machines, one of which completely wears out in a specified time and has the complete capital recovery cost assigned, and the other that doesn't wear out at all and bears only the "interest" cost.

Example 7.4:

For the dozer in Example 7.3, what would be the ownership cost if the dozer is to be used for only 3 years and then sold at its written-down value?

Answer:

Envisage the dozer as two machines: dozer A, worth $600,000 and wearing out in 3 years, plus dozer B, worth $400,000 and not wearing out at all over the 3 years.

Initial investment of dozer A	$600,000
Machine life	3 years
Capital recovery factor (see Appendix A or Equation 5.3)	0.4380
Equivalent annual payment for dozer A	$600,000 × 0.4380 = $262,800/year
Initial investment of dozer B	$400,000
Annual "interest" service for dozer B at 15%	$400,000 × 0.15 = $60,000/year
Annual payments for "complete" dozer	$322,800/year
Ownership cost (4,000 operating hours/year)	$322,800/4,000
Ownership cost for first 3 years	*$80.70/operating hour*

Example 7.4 shows a similar increase in cost per operating hour (relative to the average cost) as Example 7.2, which used the average investment method. This comes about because, over the longer life, the capital recovery factor does not correspond to paying off the capital at a constant rate. As with a house loan, the highest proportion of early payments go to servicing interest because the size of the loan is much greater in the early years. Over the original 5-year life in Example 7.3, the capital cost of the dozer is really only written down to $485,000 by the end of year 3, whereas Example 7.4 writes down the capital to $400,000 in the same time.

The equivalent lease cost method is still somewhat limited, but for quick estimates it also yields quite close results and can be adapted to extended circumstances. Nevertheless, with increased sophistication more effort is required to try to adapt the method than to undertake a discounted average cost calculation.

TABLE 7.2 Discounted average ownership cost for same case as in Example 7.1

	Year					
	0	1	2	3	4	5
"Production," *operating hours*		4,000	4,000	4,000	4,000	4,000
Initial capital cost, $	1,000,000					
Revenue at **$87.81/operating hour**, $		351,255	351,255	351,255	351,255	351,255
Depreciation allowance, $		200,000	200,000	200,000	200,000	200,000
Taxable "profit" (revenue less depreciation), $		151,255	151,255	151,255	151,255	151,255
Tax payable at 35%, $		52,939	52,939	52,939	52,939	52,939
Net cash flow, $	(1,000,000)	298,316	298,316	298,316	298,316	298,316
Discount factor at 15% return on investment	*1.0000*	*0.8696*	*0.7561*	*0.6575*	*0.5718*	*0.4972*
Present value of cash flow, $	(1,000,000)	259,405	225,569	196,147	170,563	148,316
Net present value, $	0					

Note: Numbers in parentheses indicate negative values.

DISCOUNTED AVERAGE COST

The discounted average cost method described in Chapter 5 (see "Discounted Average Cost," p. 64) can be used to determine ownership cost. This method faithfully reflects the "true" investment in an item of equipment consistent with the whole project.

The discounted average cost calculation applies to the capital components of an equipment purchase assuming operating costs are zero. In this case the goal is not to try to determine the rate per unit of *production*. Instead, an "operating hour" is the unit of production.

Example 7.5:

Using the discounted average cost method, what is the average hourly ownership cost for the same case as Example 7.1? Table 7.2 shows the calculation.

As with the discounted average cost example in Chapter 5, the calculation in Table 7.2 shows the iteratively determined result (i.e., $87.81 per operating hour) in the revenue line. The impact of 35% tax has increased the ownership cost from the previous $74.58/hour in Example 7.3 by approximately 18%.

The discounted average cost method is harder to apply than either of the previous two methods, but it has the following advantages that are difficult or impossible to obtain with the other methods:

1. The timing of capital payments can be accounted for (e.g., dragline erection, purchase costs, commissioning costs).
2. Taxation is accounted for.

3. Adjustments can be made for salvage value. This allows the equivalent capital cost of long-life equipment to be accounted for even when it is to be used just for a fraction of its technical life. The salvage value can be the transfer value to another part of the same company.
4. Irregular annual usage rates can be accounted for.

With large mining equipment, the capital component of the hourly cost is commonly more than the operating component. Estimation techniques that ignore capital run the risk of decisions being completely in error.

One significant advantage of the discounted average cost method is that it allows direct comparison of owned equipment with hired equipment. If a contractor will place a $1 million machine on hire for, say, $150/hour, then the equivalent cost of the same machine in the mining company's own hands is the operating cost per hour plus the discounted average cost as determined by this method.

CHAPTER 8 Operating Costs

Before any economic analysis or decision making can be undertaken, the operating and capital costs of equipment must be estimated. Equipment costs vary between mine sites, and there is no cost that can be applied universally.

Equipment operating costs can be developed from mine statistics, from suppliers, from contractor's quotations, and from first principles. A robust evaluation should incorporate all of these sources, with independently determined costs cross-checked with at least one other method.

The buildup of costs from statistics, suppliers, and contractors is a straightforward task that requires no further development in this text. The buildup of costs from first principles, on the other hand, is a mixture of art and science and, properly done, adds considerable value to any mine evaluation. This buildup is the subject of discussion in this chapter.

INTRODUCTION

Building up costs from first principles demands a systematic consideration of each component of equipment operation. Each component is sourced from a combination of manufacturer's formulas and historical data. When these components are aggregated, operating costs can be estimated with a high degree of confidence, especially if the factors have been calibrated from a known operation. Estimates using this method have been shown to be quite reliable even without correlation with known equipment—a case that applies to the first applications of *all* new equipment.

Operating costs change dramatically from country to country and over time. A text of this nature cannot expect to present costs that are accurate even at the time of writing. Nevertheless, some actual cases are important for illustrative purposes, and where costs are shown in this text the proportionate breakdown of costs is also shown. Even if costs change, the relative proportions of these costs may remain consistent for a longer period. Where costs are shown,

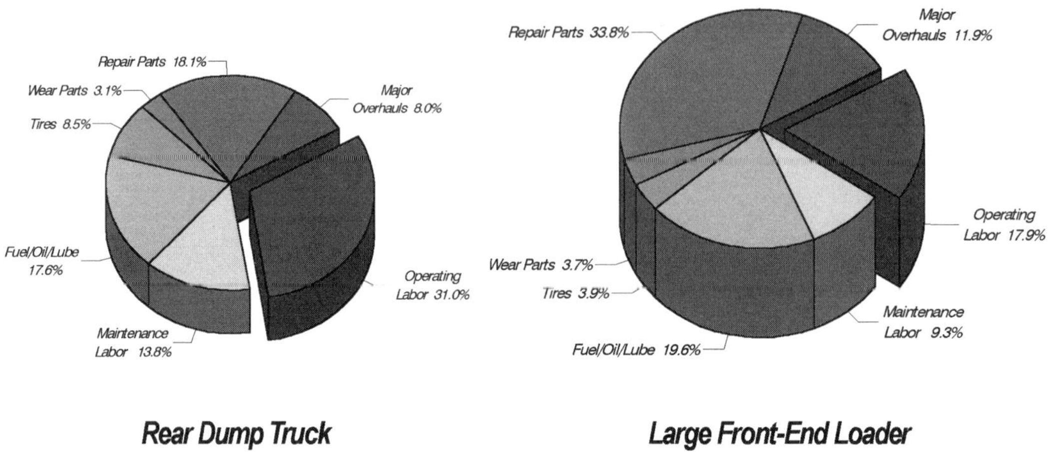

FIGURE 8.1 Typical operating costs: rear dump truck and front-end loader

TABLE 8.1 Typical operating costs: rear dump truck and front-end loader

	Large Rear Dump Truck (190 t)		Large Front-End Loader (19-m^3 Bucket)	
	U.S.$/operating hour	% of Total	U.S.$/operating hour	% of Total
Major overhauls	10.50	8.0	28.90	11.9
Repair parts	23.60	18.1	82.00	33.7
Wear parts	4.00	3.1	9.00	3.7
Tires	11.10	8.5	9.50	3.9
Fuel/oil/lube	23.00	17.6	47.50	19.6
Maintenance labor	18.00	13.8	22.50	9.3
Operating labor	40.50	30.9	43.50	17.9
Total, operating cost	*130.70*	*100*	*242.90*	*100*

they generally apply to typical open pit mines in North America or Australia at the time of this most recent update (2003).

Figure 8.1 illustrates a typical breakdown of the operating cost of a large rear dump truck and large front-end loader. As set out in Table 8.1, the standard presentation of costs is an amount *per operating hour.* Planned equipment operating hours are relatively easy to calculate, and components of operating costs are proportional to machine operating time more than to any other characteristic.

This is not the only basis for derivation from first principles. Operating costs are also a function of maximum conditions encountered, quantity of material moved or work done, and the elapse of time. Examples of these four alternative bases for buildup of operating costs are given in Table 8.2.

TABLE 8.2 Examples of alternative bases for buildup of operating costs

Example of Operating Cost Component	Estimated Primarily as a Function of:	Principally Modified by:
Fuel and oil usage	Machine operating time	Duty cycle
Tire (wear component)	Machine operating time	Abrasiveness of material
Tire (failure component)	Maximum conditions encountered, worst-case hauls	Sophistication of fleet management systems
Electricity (energy component)	Work done	Material characteristics
Electricity (demand component)	Maximum conditions encountered	Sophistication of demand management systems
Repair parts	Machine operating time	Duty cycle
Wear parts (bits, ropes, teeth)	Quantity of material moved, or work done	Material characteristics
Some components of major overhauls (e.g., painting)	Elapse of time	Environmental conditions

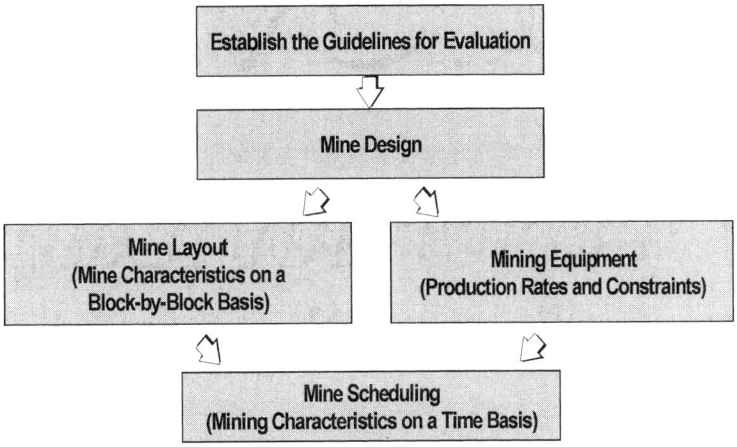

FIGURE 8.2 Flowchart to prepare mine schedule

OPERATING SCHEDULES

Operating costs are derived mainly from machine operating hours, but what is the source of the machine operating hours?

Chapter 3 set out the systematic planning process that yields the operating schedules prior to costing. For any one level of detail (each iteration in the planning cycle), the steps are as shown in Figure 8.2. The mine design establishes the direction of mining, the position of access points to the mine working areas, and certain other constraints on mining, but it does not fix the sequence of mining, the rate of mining, or even the equipment to be used. Although the *process* of laying out the mine takes cognizance of the equipment to be used (e.g., mining blocks must be wide enough to allow equipment to operate), its primary purpose is to calculate the quantity and quality characteristics of ore and waste on a *block-by-block* basis. The task of laying out the mine is undertaken concurrently with the task of determining production rates and constraints on equipment usage. Equipment production rates may be a function of face height and ore type (and so, during the scheduling process, may vary from block to block), but they can also be a function of time. Winter production rates may be quite different than summer production rates, for example.

It is only when equipment (production rates and constraints) is combined with the mine layout in the *scheduling* phase of the process that quantities and quality characteristics are available on a *time* basis.

The mine-scheduling task is analytically described in four steps:

Operation ...	Using ...	Expressed in ...
1. Start with ...	A mining block, with a quantity of ore or waste	Tons or cubic meters
2. Divide by ...	Production rate of equipment	Tons per operating hour
3. To determine ...	Elapsed time to complete task	Machine operating hours
4. Tabulate ...	All operations (quantities moved, meters advanced, machine operating time applied, supporting requirements) between each starting and ending point in time (month to month, year to year, etc.)	

The output from the mine-scheduling phase is a set of tables with information on a month-by-month, year-by-year, or other time basis. Information includes mining quantities, operating requirements (consumables such as rock bolts, explosives), and machine requirements (operating hours and numbers of machines). Tables 8.3 and 8.4 show two simplified schedules of this form.

Similar schedules are prepared for workforce numbers categorized by task and for equipment purchase and replacement requirements. A more comprehensive description of these schedules for a whole-mine study is given in Chapter 11.

SOURCES OF COST INFORMATION

There is no single simple and reliable source for estimating mining equipment operating costs. The recommended approach is to develop costs from first principles and then to cross-check from actual mine statistics and by whatever rules of thumb are available. A number of rules of thumb are listed in this chapter. Spreadsheets and databases allow data to be stored, correlated, and updated quickly.

The advantages and importance of calculating costs from first principles and *then* cross-checking them are as follows:

- Costs can be developed for any equipment, not just equipment already in use on the mine site.
- There is consistency between costs (see Example 8.1 at the end of this list).
- There is less probability of a "day 1" deviation. The equipment may have been operating with abnormally high or low costs from day 1, but the company would not necessarily know this from its own statistics. Costs derived from first principles allow more ready comparison with outside operations.

TABLE 8.3 Example schedule: mine operating requirements

Operating Cost Item	Units	Year 1	Year 2	Year 3
Explosives: ANFO	tons	2,500	3,600	4,000
Explosives: slurries	tons	600	770	800
Reclamation: fertilizer	tons	—	35	120
Reclamation: seed	kilograms	etc.	etc.	etc.
⋮	⋮	⋮	⋮	⋮
Other consumables			*Amount consumed per year*	

TABLE 8.4 Example schedule: major equipment operating hours

Equipment Item	Units	Year 1	Year 2	Year 3
Model A rope shovel	Operating hours	4,000	6,500	6,500
Model B rear dump trucks	Operating hours	16,300	27,800	29,200
Model C front-end loader	Operating hours	5,000	6,000	6,000
Model D rear dump truck	Operating hours	etc.	etc.	etc.
⋮	⋮	⋮	⋮	⋮
Other equipment			*Total fleet operating time per year*	

Example 8.1:

If you are using only mine-site statistics, and at your site your smaller loader always works on weathered overburden whereas your large loader always works on basalt, then the operating cost estimates between loaders at your mine site are not comparable. The harsher environment under which the large loader is working will probably lead you to conclude that the smaller loader is more reliable and/or has a lower comparative cost when in fact the reverse may be the case. This problem also occurs with equipment of different ages.

Typical sources of data are shown in Table 8.5.

It is sound practice to cross-reference sources of data to highlight any discrepancies. Systematic derivation from first principles allows consistent comparisons across different machines. It also allows easy aggregation of the same common elements across different machines, e.g., fuel storage requirements and total maintenance staffing.

In the process of obtaining cost estimates, it is critical to understand exactly what is and what is not included. For example, production rates *per operating hour* may use a different definition than operating costs *per operating hour.* The mining equipment supply, mining contracting, and mining consulting businesses are all very competitive, and in this environment firm quotations from any of these sources seldom differ by substantial amounts. New technology is always reducing the costs and increasing the availability of equipment, but, absent any dramatic breakthrough, any costs that appear to be substantially

TABLE 8.5 Sources of cost information

Source	Comment
Operating mines with similar plant	Definitions and cost allocations vary from company to company. Usage data (e.g., liters of fuel used per hour, tire life) are more reliable from mine to mine than are direct cost data (e.g., fuel cost per operating hour, tire cost per operating hour).
Manufacturers	Cost data may not be consistent from manufacturer to manufacturer but should be reliable across machines from the same manufacturer. Manufacturers are a good source of data for regular repair and overhaul costing, but they are generally less knowledgeable than mine sites for wear items (life of ropes, tires, etc.) and maintenance due to breakdowns and accidents. Manufacturers will sometimes provide contracted (fixed) prices for operating costs.
Consultants, industry cost services	Consultants are adept at mine-to-mine comparisons and at identifying anomalies in components of cost. Consultant costs are probably the most appropriate source with new equipment and in new applications of existing equipment.
Government and industry authorities	These sources are appropriate for broad indications of costs. The standardized forms used to collect data are sometimes inappropriate for many of the mines obliged to supply data, and the type of information is frequently dated.
A company's own operations	Internal costs are valuable sources of information. Past history may not be a reliable guide if the new application differs from the past. Some cost-accounting systems use fairly arbitrary rules for allocation of costs not *specifically* collected on individual items of equipment.
Contractor quotations	For such items as explosives, tires, etc., these sources provide the most reliable costs. Quotations should differentiate between (1) "budget" estimates that typically mean some standard list price and (2) costs that might ensue after a competitive bidding process.
Rules and formulas	Rules and formulas such as described in this chapter and in manufacturer handbooks provide very reliable estimates once enough studies have been cross-checked against actual costs.

different (lower) than costs from alternative sources should always be questioned and treated with caution.

SUPPLY COSTS

Supply costs include fuel and lube, electrical costs, tires, and explosives. Some typical examples are

- fuel = $0.27/l ($1.02/gal)
- power = $0.04/(kW-h) for energy and $10.12/kVA per month for demand
- greases = $1.60/kg ($0.73/lb)

These basic costs are necessary inputs to calculate hourly equipment costs, but they need to be checked for individual mine sites.

JOB CONDITIONS

Job conditions have a direct effect on costs. Costs are first estimated for typical or average conditions and are then adjusted up or down depending on the degree of difficulty of the excavation and the nature of the material.

TABLE 8.6 Job conditions

Condition	Description
Good conditions (easy digging)	Material is relatively loose and free-flowing. Equipment operates with considerable idling or low power. Long life of wear items can be expected because of low abrasiveness. Low digging power is required, and material heaps well into the bucket. Tires wear out rather than fail because of cuts and abrasion. An example indicating good conditions would be a dozer that is required to work only part-time on a coal stockpile.
Average conditions	Material requires blasting to maintain productivity. Some power is required to penetrate the bank, and material heaps reasonably well. Engine has periods of full power but still some idle periods. Wear rates are moderate. An example indicating average conditions would be a shovel loading well-blasted shale.
Poor conditions (difficult digging)	Higher powder factors are required, and often the material is bulky, is irregular in shape, and has poor fill factors. Engine is often at full power. Tires fail because of rock cuts and abrasions. Wear rates are high and component life is reduced. An example indicating poor conditions would be a contractor dozer ripping strong sandstone.

Example 8.2:

A mine has two identical dozers. One works on the coal stockpile and has 80% of the average cost. The other dozer is assigned to ripping and pushing partings and has 130% of the average cost. The initial purchase price is the same.

Table 8.6 shows a typical classification of job conditions. Other factors affecting costs and productivity include (1) such material characteristics as abrasiveness, bulk density, flow properties (sticky versus free-flowing), strength, degree of blasting and fragmentation, and joint spacing; (2) such labor factors as operator skills, the cost of labor, and the overall management conditions; (3) the proximity of spare parts and maintenance support such as cranes; and (4) such utilization factors as annual working hours and the time allocated for service and overhauls.

OPERATING COST DATA

This section covers operating costs for power (energy and demand), fuel and lubrication, tires, maintenance supplies (repair parts), operating supplies (wear parts), major overhauls, and labor costs.

Power (Energy and Demand)

Operating costs associated with electrically powered equipment include a charge for the use of the energy component, as well as a charge reflecting the required installed capacity of the power-generating facility. This demand charge comes about because of the cyclical loads of most mining machines. The electric utility has to be able to supply high power for short periods of time, yet on energy charges alone the utility would be paid for only a much lower actual average usage.

In the case of household electricity, the utility company bills the householder a rate based only on the kilowatt-hours used. This is an overall charge calculated to pay for the utility's supply costs (its fuel supply, labor, etc.) as well as

TABLE 8.7 Electricity (energy) usage for selected set of mining equipment

Equipment	Electricity (Energy) Usage
Rope shovel (older, smaller shovels)	0.6 kW per m^3/h (0.45 kW per yd^3/h)
Rope shovel (newer, larger shovels)	0.35 kW per m^3/h (0.27 kW per yd^3/h)
Walking dragline	1.5 kW per m^3/h (1.15 kW per yd^3/h)

its "capital" costs. Over a whole city, there is a lot of diversity among domestic consumers, and electric utilities can predict load demands reliably.

In a large mine, large demands can sometimes be applied for short periods during the start-up of large mining machines, and prediction of when this might occur is difficult. Rather than try to predict the demand (and bill the mining company on the basis of an overall rate), most utilities bill their industrial customers separately for the energy used and for the demand. The energy charge is meant to reflect the utility's operating costs, and the demand charge is meant to reflect the utility's capital costs. Demand charges are normally established in relation to the maximum 15-minute electricity demand occurring within a month.

The energy component of the charge is best determined in proportion to the amount of work performed. For quick estimates (i.e., a rule of thumb), the average power for a selected set of mining equipment is as listed in Table 8.7.

Since most mining machinery (particularly shovels and draglines) completes a typical excavation cycle in less than 1 minute, it is not the cyclical operation that has an impact on the demand over a 15-minute period. Higher-than-average demands occur because of irregularities in the operation (such as start-up) and variability in digging conditions as the machine goes about excavating its block of dirt.

For planning purposes a mining company can estimate demand (maximum kilovolt-amperes) in advance based on the expected average power (energy) usage. For cyclical machines or machines with variable duty cycles, such as crushers, demand (in kilovolt-amperes) is typically 10–15% higher than average power (in kilowatts). For machines operating constantly, such as pumps and generators, demand (in kilovolt-amperes) is equal to average power (in kilowatts). The demand charge is a function of the maximum conditions encountered (in a month, usually), so this charge will be incurred whether the machine works 1 hour per month or 600 hours per month. If back-allocated to machine operating hours, this scheduling influence makes a dramatic difference. The proportion of the electricity bill attributable to demand charges may be much greater than for energy charges in cases of low utilization.

Example 8.3:

A medium-sized modern rope shovel is the only electrically powered machine on-site. It operates on shift schedules ranging from 200 hours per month (day shift only, during summer) up to 600 hours per month (three shifts per day during winter) at a production rate of 1,200 m^3/h. Power costs \$0.04/(kW-h) and demand charges are \$15.00/kVA per month.

Estimated energy usage	0.4 kW per m^3/h (from Table 8.7)
Average hourly electricity (energy) usage	0.4 kW per m^3/h × 1,200 m^3/h = 480 kW
Hourly energy cost	Average (hourly) energy usage × hourly energy charge = 480 kW × \$0.04/(kW-h) = \$19.20/operating hour

The monthly demand charge may be calculated as

$$
\begin{aligned}
\text{monthly demand charge} &= \text{average hourly usage} \times \text{factor for variability} \times \\
&\quad\ \text{demand charge} \\
&= 480\ \text{kW} \times 1.15 \times \$15/\text{kVA per month} \\
&= \$8{,}280/\text{month}
\end{aligned}
$$

This value is fixed regardless of the number of operating hours. Different levels of utilization have the following cost breakdowns:

Cost Component	Low Utilization (200 hours/month)	Medium Utilization (400 hours/month)	High Utilization (600 hours/month)
Energy cost, *\$/operating hour*	19.20	19.20	19.20
Demand charge, *\$/operating hour*	8,280/200 = 41.40	8,280/400 = 20.70	8,280/600 = 13.80
Total electricity cost, *\$/operating hour*	60.60	39.90	33.00

Demand charges are frequently allocated to individual items of equipment for costing purposes in a mine-planning study, but the fixed element in the demand charge means that the hourly cost is sensitive to the schedule of operation. Demand charges are seldom allocated to individual equipment in mine cost accounts.

The situation becomes more complicated when there are inconsistencies in the operating schedules of equipment, giving periods of irregular high demand. If, for example, overburden equipment works three shifts per day, 7 days per week, but other production equipment works only day shift, then the demand should be calculated by working out the average number of kilowatts used in a typical hour *during the heaviest demand period* and then applying the 10–15% increase suggested for the highest-production 15-minute period in the month.

If a large item of electrical equipment (e.g., a dragline or a mill) has to be started and stopped, the demand over a 15-minute period during start-up will probably exceed the demand during normal operation. This demand will then determine the number of kilowatts to apply to the demand charge.

This may be particularly important in the case of underutilized draglines and other large equipment. A dragline that works 6,500 hours per year may incur demand charges averaging, say, $100 per hour, but the same machine worked on day shift, 5 days per week (totaling 1,600 hours/year), and started and stopped each day will incur charges averaging $390 per hour. It may be possible to renegotiate charges with some utilities if changes in schedule result in such dramatic changes in cost.

It is not uncommon in mines using a lot of cyclical, electrical machinery to find that the demand charges exceed the energy charges. On its own this cost dissection is not of any significance. Recall that demand charges are aimed at recovery of capital costs, whereas energy charges are aimed at recovery of operating costs. The "capital" or "ownership" cost of running a dragline or shovel also exceeds the operating cost, and power-generating facilities are similar, highly capital-intensive operations.

Fuel and Lubrication

Fuel costs are based on

- the cost of fuel
- the engine's fuel consumption rate, which depends on engine power and the duty cycle
- working conditions

A truck hauling on a long, loaded, uphill haul will have a higher duty cycle than a truck hauling on a short, flat haul. Both trucks would have duty cycles less than 50% (typically) because they are using only small amounts of power during the return cycle and when they are being loaded. Diesel-powered pumps and lighting plants can have near 100% duty cycle. Diesel-powered shovels and draglines *may* have slight variations in usage with duty, but this is not likely to be significant for cost-estimating purposes.

Fuel consumption for most large diesel motors working at 100% load factor is approximately 0.3 l/kW per hour (0.06 gal/bhp per hour, or 0.35 lb/bhp per hour). Load factors range from 0.2 to 0.8. The formula for this calculation is

$$\text{hourly usage (liters)} = \text{rated power (kilowatts)} \times 0.3\ \text{l/kW per hour} \times \text{load factor}$$

or

$$\text{hourly usage (gallons)} = \text{rated power (brake horsepower)} \times 0.06\ \text{gal/bhp per hour} \times \text{load factor}$$

For example, a truck with a 1,300-kW (1,743-hp) engine hauling under average conditions (load factor 0.35) has a fuel consumption of 1,300 × 0.3 × 0.35 = 137 l/operating hour (36.2 gal/operating hour).

Tables of the load factors or actual consumption for common diesel-powered mobile equipment are available from manufacturers' handbooks. Table 8.8 shows some common load factors back-calculated from a handbook for an example set of equipment. As the table shows, graders and trucks typically

TABLE 8.8 Load factors for fuel usage calculation

Equipment	Power (kW)	Load Factor, Low Range	Load Factor, High Range
Tracked dozers	160	0.40–0.52	0.67–0.83
	276	0.36–0.51	0.63–0.83
	575	0.36–0.41	0.63–0.67
Wheel-dozer	336	0.40–0.45	0.71–0.77
Grader	205	0.31–0.41	0.62–0.72
Hydraulic excavator	287	0.30–0.35	0.69–0.74
Scrapers	366	0.36–0.41	0.66–0.71
	443	0.35–0.43	0.65–0.71
	708	0.41–0.46	0.72–0.77
Rear dump trucks	485	0.18–0.26	0.38–0.49
	649	0.18–0.27	0.38–0.50
	962	0.18–0.28	0.35–0.50
	1,272	0.18–0.27	0.37–0.49
	1,534	0.18–0.26	0.37–0.49
Front-end loaders	280	0.38–0.45	0.71–0.79
	515	0.35–0.39	0.67–0.73
	932	0.36–0.39	0.68–0.74

Note: 1 kW = 1.34 bhp

have lower load factors than loaders and dozers, but from one size of machine to another the variation is not substantial. Applying load factors to engine sizes and *then* multiplying by the price of fuel is a more reliable method of estimating fuel costs than simply tracking fuel costs by equipment. This method is more robust as the price of fuel changes over time and across countries. It also provides an automatic adjustment mechanism when manufacturers upgrade engines on their machines.

Lubrication costs can usually be calculated as a percentage of the hourly fuel *costs.* These proportions range from 20% for equipment with a relatively low proportion of hydraulic components (such as a tractor-trailer coal hauler) up to 30 to 40% for equipment with a high proportion of hydraulic components (such as a hydraulic excavator). Adjustments plus or minus 5 percentage points may be made to these figures depending on how severe the duty cycle is.

Alternatively, the consumption rate can be expressed as either fuel volume (liters or gallons) per hour or fuel weight (kilograms or pounds) per hour. Values can be obtained from the equipment manufacturer or from operational records. These values are then multiplied by their appropriate unit cost. This is a more accurate method and is about the only method for large electrically powered equipment that consumes substantial quantities of lubricants but no fuel oil. Since lubrication changeout periods are typically well understood and adhered to, calculation of lubrication cost by this method, though tedious, yields very reliable results.

Tires

This section does not apply for tracked or tub-mounted equipment, such as track dozers, shovels, and draglines, but it is directly applicable for rubber-tired equipment, such as trucks, front-end loaders, rubber-tired dozers, graders, and light vehicles.

Tire costs are obtained by the total cost of the tire multiplied by the number of tires and divided by the hourly life of the tire. Tire manufacturers provide guidelines for calculating hourly life. This is usually a base number of hours (4,000 is a common base) multiplied by a series of factors. The factors adjust for the conditions shown in Table 8.9.

The base number of hours is multiplied by the series of factors to give the total life. In coal mines the tire life can vary from 1,500 hours to 12,000 hours. As a rule of thumb, the general average for coal haulers in a coal mine is around 5,000 hours. In the worst conditions (e.g., a wet quarry, mining abrasive basalt, tires not equipped with chains), tire life for a front-end loader may be as little as 600 hours. Tire life for rubber-tired dozers is usually more dependent on damage than hourly usage.

Example 8.4:

A truck has six tires costing $15,000 each. The estimated life is 4,000 hours. The tire cost per hour is

$$\begin{aligned}\text{tire cost per hour} &= \frac{6 \text{ tires} \times \$15{,}000/\text{tire}}{4{,}000 \text{ hours/tire}} \\ &= 22.50/\text{tire}\end{aligned}$$

Low tire life is usually the result of failure of the tire rather than the tire physically wearing out. In mines located in tropical areas, excessive heat buildup has traditionally been the underlying cause of such premature failure, but the increasing use of radial ply tires has helped resolve some of these problems. Other failures are the result of damage, misuse, deficient product, unsuitable specification, or poor road surface. In hot, metalliferous mines, it is not unusual for 30% of tires to fail prematurely as a result of these sorts of causes.

TABLE 8.9 Conditions impacting tire life

Condition	Comment
Tire maintenance conditions	Small cuts and tire damage can often be repaired at low cost if recognized quickly.
Speed	Heat buildup in tires is proportional to speed, leading to premature failure if maximum tire rating is exceeded; long, slow hauls and time spent under the loader reduce average speeds.
Tire loads and amount of overloading	Heat buildup in tires is proportional to tire load; loaded hauls downhill (with empty return uphill) increases average tire loading.
Surface conditions, including temperatures	Poorly maintained roads lead to tire damage; high ambient temperatures can exacerbate heat buildup.
Wheel conditions	Front-wheel maneuvering and rocks caught between dual wheels impact tire wear and damage.
Number of curves and grades; asymmetric loading	These increase tire flexing and potential for tire failure—particularly if tire loads are near rated maximum.

Maintenance Supplies (Repair Parts)

The cost of repair parts is one of the most difficult to calculate. If no historical data are available, there are two commonly used methods of estimation.

The first general formula, which is appropriate for large equipment, such as shovels, draglines, and crushing/conveying systems, involves multiplying the capital cost by a percentage and dividing this result by the number of operating hours per year. Typical values range from 3% (for a conveying system consisting mostly of structural work) to 10% (for a bucket-wheel excavator system, which has a much higher proportion of "machinery" and moving parts). Appropriate allowances must also be made for digging conditions. Most manufacturers will assist with developing maintenance and cost programs for this sort of large equipment.

A second method uses an hourly repair factor or repair parts factor. This method, which is widely used by equipment manufacturers for smaller equipment, assumes that equipment is a collection of spare parts. Under this assumption, the capital cost of the machine is the cost of this collection of parts less a discount for buying in bulk. Spare parts costs can be estimated as a proportion of the initial capital cost. Some of these parts last 500 hours, while some last over 10,000 hours. Starting with a selected benchmark, or reference, operating life, one can calculate the total cost of parts expected to be purchased throughout this benchmark time. Dividing the total cost of parts during this period by the life of each part yields the hourly cost of parts. By historical convention the first 10,000 hours of machine life are used for this reference life. This method does not assume that the equipment lasts only 10,000 hours—the figure of 10,000 hours is selected only as a reference life.

As with the load factor calculation, there is a lot of consistency between types of equipment. For example, if a dozer costing $1 million typically has $200,000 (i.e., 20% of initial cost) of repair parts in the first 10,000 hours of life, then a dozer costing $500,000 will typically require $100,000 of repair parts in its first 10,000 hours of life. Front-end loaders may incur repair parts costs equivalent to 25% of their initial capital cost, but this percentage will be quite consistent over a whole range of loader sizes.

Although this method of estimation is widely used, its success stems from a historical era that is less applicable now. Three limitations are important to recognize:

1. It estimates costs over the whole life, but in practice these costs are less in the early years of machine life, and they increase as the equipment gets older.
2. The commonly used reference life of 10,000 machine operating hours started in an era when a lot of equipment lasted for only 10,000 hours. For 20,000 hours of life, an extrapolation of costs (using *extended life* multipliers) still yields reliable results. For 40,000 hours or more of machine life (now commonplace with mining machinery), reliable extrapolation from a 10,000-hour starting point is questionable.

3. Planned maintenance periods and component life schedules rarely have any allowance for damage due to accidents.

There are two additional problems that follow from the "averaging" procedure adopted with this estimating method:

1. Economic analysis requires costs *as they occur* for cash flow tabulation. The use of average costs overestimates cash costs in the early years of equipment life and underestimates cash costs in later years.
2. If used for economic analysis of equipment replacement strategies, the method yields erroneous results. It biases the comparison in favor of older equipment and *against* replacement.

After the shortcomings are accounted for, both of the methods just described have one important characteristic—they are relatively objective in their derivation of results. The alternative—using personnel familiar with a limited range of equipment to make subjective judgments—is substantially less objective. All methods require judgment in applying repair factors and job condition factors.

The estimating process is complicated by significant trends in mining machinery design. With the widespread use of computer aided design, component sizing can be finely tuned, and components can be designed for replacement (to wear out) at consistent rates. Thus, it is more economical to replace entire units (e.g., the complete final drive assembly) than to replace components individually. This means that, compared to experience prior to the 1990s:

- Whole parts are replaced rather than repaired. The proportion of maintenance labor to maintenance parts continues to decrease.
- When whole parts are changed out, it is more efficient to return them to the factory or an off-site facility. The number of skilled site personnel engaged in maintenance repair continues to decrease.
- Because of more efficient design, modern machines have lower cost and less wastage on repair parts compared to previous designs of the same type of machine of similar age. The problem where one part is worn out but its adjacent part is only 25% worn (but replaced anyway) is becoming less of an issue.

As with any other estimating technique, additional allowances must also be made for specific site conditions. Trucks driving on well-made roads will be subject to less stress and longer component life than trucks driving on rutted, poorly maintained roads. Other equipment is similarly affected by job conditions. Allowances for specific job conditions typically range from 0.8 (80% of "standard" repair parts usage) for well-maintained and -managed operating environments to 1.2 (120% of "standard" repair parts usage) for poorly maintained and poorly managed operating environments.

Example 8.5 shows the standard calculation of repair parts applying the factors and related method described earlier.

Example 8.5:

A dozer with an initial capital cost of $1 million is working under good job conditions and is to be used over a 15,000-hour life. (The "standard," or reference, life of this machine, to which the repair factor applies, is 10,000 hours.)

Capital cost	\$1,000,000
Hourly repair factor	0.2 (based on 10,000 hours)
Adjustment for job conditions	0.8
Repair cost, first 10,000 hour life	$\frac{\$1{,}000{,}000 \times 0.2 \times 0.8}{10{,}000} = \$16.00\text{/hour}$

In this case, the dozer is expected to last 15,000 hours, and (according to the manufacturer handbook) the average repair parts usage over 15,000 hours is 10% higher than the average over the first 10,000 hours.

Extended life multiplier	1.1
Repair cost, "planned" life	$16.00 × 1.1 = $17.60/hour

The extended life multiplier applies to the *whole* of the life. If one wants the machine to last the longer (i.e., 15,000-hour) life, then one must expect 10% percent higher repair costs *on average* throughout the whole of this life. Some of these higher costs would be incurred during the initial 10,000 hours of life and some during the extended life.

Table 8.10 sets out a series of repair factors for common mining equipment. This table does not use an "extended life" or extended life multiplier; rather, it supplies just one factor for the typical life shown. Example 8.6 shows a sample calculation using data from this table.

Example 8.6:

A large rear dump truck working in a coal-mining application under good conditions has a current purchase price of $1,900,000, including $120,000 for the initial set of six tires. The net capital cost is thus $1,900,000 – $120,000 = $1,780,000. From Table 8.10 the repair factor is 0.25×10^{-4}, and for good conditions the job factor is 0.8. The estimated repair cost can be determined as follows:

$$\begin{aligned}\text{estimated repair cost} &= 1{,}780{,}000 \times (0.25 \times 10^{-4}) \times 0.8 \\ &= \$35.60/\text{operating hour (average over whole life)}\end{aligned}$$

From Table 8.10, the major overhaul cost is estimated at 15% of $1,780,000 every 15,000 hours:

$$\begin{aligned}\text{major overhaul cost} &= (0.15 \times 1{,}780{,}000)/15{,}000 \\ &= \$17.80/\text{operating hour}\end{aligned}$$

TABLE 8.10 Typical repair parts and maintenance labor factors

Typical Equipment Item	Typical Life (operating hours)	Hourly Repair Factor, Typical Life	Major Overhaul (% of capital)	Frequency of Major Overhaul (operating hours)	Maintenance Labor (person-hours per operating hour)
Dozer	22,500	0.25×10^{-4}	15	10,000	0.4–0.7
Walking dragline	100,000	0.035×10^{-4}	3	20,000	1.7–3.1
Coal drill	35,000	0.25×10^{-4}	12.5	10,000	1.1–1.5
Overburden drill					
Electric	75,000	0.15×10^{-4}	10	15,000	1.1–1.7
Diesel-hydraulic	40,000	0.25×10^{-4}	12.5	10,000	1.1–1.5
Grader	22,500	0.25×10^{-4}	15	10,000	0.3–0.5
Front-end loader	30,000	0.30×10^{-4}	15	10,000	0.5–0.8
Hydraulic excavator	27,500	0.25×10^{-4}	15	10,000	1.1–1.5
Rope shovel					
Hard rock	80,000	0.075×10^{-4}	17.5	20,000	1.2–1.5
Coal	100,000	0.035×10^{-4}	7.5	20,000	1.0–1.25
Bottom dump truck	40,000	0.25×10^{-4}	15	15,000	0.7–0.8
Rear dump truck	45,000	0.25×10^{-4}	15	15,000	0.5–0.8

Also from Table 8.10, the maintenance labor requirement is estimated at 0.7 person-hours per operating hour. The (assumed) labor cost is \$39.50/person-hour (see the "Labor Costs" section later in this chapter):

$$\begin{aligned}\text{maintenance labor cost} &= 0.7 \times 39.50 \\ &= \$27.65/\text{operating hour}\end{aligned}$$

The data in Table 8.10 change considerably over time. The most significant changes noticed in the past 15 years are as follows:

- The typical life of mobile equipment has increased. In the 1970s, large haul trucks were typically installed with a planned life of 20,000 hours or 5 years. With better design, equipment monitoring, component changeout, and longer annual usage, typical life has extended to 8 years or more, and 40,000 hours or more. The same trend is evident with dozers, loaders, hydraulic excavators, and other mobile equipment.
- The typical life of less-mobile large equipment, such as draglines and rope shovels, has reduced. There are many shovels and draglines in use around the world that are 30 or more years of age, yet an increasing trend is for equipment to be superseded (and often, stood down) after 12 years or more life. This trend reflects advancing technology of new draglines and shovels (a change in *economic* life) rather than any physical change.
- Maintenance labor costs show a continual decline resulting from increasing component changeout and decreasing on-site repair. Component life can be matched to planned changeout times, with efficient planned maintenance periods replacing though not eliminating ad hoc day-to-day breakdown maintenance.
- The capital cost of equipment has reduced substantially (in inflation-adjusted terms), and the repairs and major overhauls are estimated in proportion to this capital cost. This has resulted in a commensurate reduction

in repair parts and major overhaul costs. Some of the most significant reductions in capital cost (and therefore maintenance cost) have occurred in the larger machinery (shovels and draglines). Reductions have come from increased competition and through computer aided design technologies that allow manufacturing efficiencies previously available only to mass-volume machines.

- An increasing trend is for outsourcing of maintenance, often to firms that are associated with the original equipment manufacturer. Because of their focus on equipment maintenance only, these firms enjoy efficiencies and reduced costs that are unavailable to mining companies undertaking the same tasks at their own mines. If the work is undertaken by their own personnel operating under workshop conditions, manufacturers typically quote maintenance labor requirements that are only 50–70% of the values shown in the last column of Table 8.10.

Operating Supplies (Wear Parts)

Wear items include bucket teeth, hoist ropes, drag ropes, and dump ropes for draglines and shovels; ripper boots, grouser plates, and cutting edges for dozers; bits, adapters, and drill stems for drills; and wear plates on conveying and crushing machinery. Sometimes referred to as "ground-engaging tools," these operating supplies (consumables) are usually estimated and costed separately from regular maintenance parts costs.

There is no universal method for estimating these costs quickly. Manufacturer experience or the experience of other mines operating in the area has to be used (though it may not be directly applicable).

For most open cut mines, wear parts are normally only a small proportion of the operating cost of equipment, and errors in their estimation don't translate into large errors in overall cost estimation.

Underground mines are not as fortunate from this perspective. In metalliferous mines drilling is a significant part of the cost structure, and drilling is quite sensitive to material conditions and wear parts costs. The productivity and cost of underground coal-mining equipment are equally sensitive to material conditions, and replacement costs of picks and of wear plates on continuous miners, shearers, and conveyors are significant.

The method of calculating wear parts cost is to take the cost of all of the wear items and divide each of them by its estimated life. For example, if a ripper boot costs $400 and has a life of 25 hours, the cost is 400/25 = $16/hour. Some typical wear items on a number of common mining machines, with values of possible life, are set out in Table 8.11. Some wear items (for example, electrical brushes) for which the life is not dependent on material conditions are commonly categorized as part of normal maintenance costs.

TABLE 8.11 Typical equipment wear items and possible life values

Item	Comment
Hoist ropes (on a dragline)	May last 4 to 6 months. Will normally be end-for-ended after perhaps 3 months.
Hoist ropes (on a rope shovel)	Gold mines or hard-rock mines: typically 500 to 800 hours Coal mines, less abrasive applications: typically 1,200+ hours
Bucket teeth (on a dragline or rope shovel)	May last 4 hours (in hard or extremely abrasive material) to 4 weeks (in soft, easily dug material). Usually turned over when part worn.
Bucket adapters, shrouds	Rebuilt along with bucket rebuilding. Undertaken perhaps monthly (dragline buckets changed over, shovel dippers repaired on the machine).
Track wear plates (on dozers)	Depends on abrasiveness of ground and importance of traction. For ripping tasks, may last only 1 to 3 weeks; for coal stockpile work, may last 1 year.

TABLE 8.12 Common tasks during major equipment overhauls

Item	Common Tasks During Major Overhaul
Draglines	Gear-case realignments; tub strengthening and replacement of tub wear plates; tub roller circle realignments, replacement; boom sheave bearing replacement; structural repairs to the boom; repainting
Shovels	Replacement, realignment, and repair of sticks and arms, dippers; refurbishment of motors, gear-boxes; replacement of tracks and propel mechanisms; updating of electrical controls
Hydraulic shovels	Replacement or major refurbishment of engines, hydraulic pumps, tracks, and rams
Trucks	Replacement or major refurbishment of engines, converters, transmissions, final drives, and brakes; frame realignment and rebuild; tray rebuild or replace

Major Overhauls

Major overhauls cover the cost of major component exchange or rebuild. This can be estimated as a percentage of initial capital cost (such as 15% every 12,000 hours) or else as a buildup of components and their lives. For large semimobile equipment such as draglines, drills, and shovels major overhauls are commonly scheduled every 5 years.

Table 8.10 gives the typical frequency of major overhaul for a range of common mining equipment. "Maintenance" items commonly undertaken during major overhauls are set out in Table 8.12. These items are estimated in a similar manner as repair parts. See Example 8.6 for a sample calculation.

Labor Costs

Labor costs are typically calculated on an annual basis. Subsequently they are broken down into a rate per hour (for exercises involving equipment operating costs) or a rate per week.

Operating Labor For a whole-mine study, operating labor requirements are usually built up from two components: (1) labor directly associated with equipment and (2) labor that is a function of the task being undertaken. In open pit mining, almost all operating labor is associated with equipment, whereas in underground mining most operating labor is a function of the task.

Operator labor costs are developed in proportion to machine operating time, or from time spent "at the face." The following factors are considered:

- The shift roster. Shift rosters vary from mine to mine. The archetypal roster for 5-day-per-week operation is a three-panel roster with day, afternoon, and night shifts. The crew on day shift one week will start on afternoon shift the following week. A common continuous shift roster (7-day-per-week operations) involves four crews covering the complete 168 hours in a week. Each crew works four shifts each of 12 hours duration before being rostered off for an average of 3 days. In this roster, two crews are always "on duty," and two crews are always "rostered off," with shift schedules cycling over a 4-week period.
- Industrial practices. Mines in some parts of the world require two operators on equipment even though the machine has been designed for operation by a single operator. Remote activities during night shift sometimes require two persons for safety reasons.
- Absenteeism and availability of personnel to cover periods of annual leave, sickness, and training.
- Availability of equipment. When mobile equipment is unavailable, it does not normally have an operator assigned to it. In contrast, when large fixed or semimobile equipment is unavailable (broken down), operators are often still required to assist with maintenance or to undertake other duties such as cleanup.

Rope shovels and draglines in coal mines are typically staffed continually, even when they are on maintenance or broken down. The numbers of rope shovels and large electric drills in metalliferous mines may be determined by grade control requirements, not production requirements, and only 50% of them may be staffed at any one time. Conversely, if there are 20 trucks in the fleet and the expected availability is 80%, then normally only 16 trucks are staffed. A pool of personnel or personnel reallocated from less-urgent tasks may be used for unplanned absenteeism.

Regular overtime (or a roster that schedules personnel for 50 or more hours per week) is often a cost effective way of reducing the total number of personnel on-site. Though this commonly means higher hourly costs for the same production, overall costs may be less. This is particularly the case for remote mine sites where each employee is accompanied by high fixed costs, such as company supplied accommodation. The trade-off between additional *scheduled* overtime or additional personnel should be considered. As with equipment productivity and operating cost, the number of person-hours assignable to a single employee per year must be consistent with the calculated cost per employee per year. Employees scheduled to work overtime on a regular basis obviously cost more per year than employees assigned to the same task for a lesser number of hours each year.

Example 8.7:

A conveyor-based excavating system needs two operators at any one time. The system works three shifts per day, 5 days per week; after allowance for availability, it operates for 3,000 net operating hours per year. Each operator costs \$60,000 per year. What is the hourly labor cost? Assume personnel are still required on service days. Absenteeism runs at 8%.

$$\text{operating labor cost} = \frac{3 \times 2 \times \$60{,}000}{3{,}000 \times 0.92}$$
$$= \$130.43/\text{operating hour}$$

Maintenance Labor There is no universal method to estimate maintenance labor requirements. Factors to allow for are

- how much work is done "off-site," such as component exchange
- nature of the operation (hard digging or not)
- skill and experience of operators and maintenance personnel
- proximity of spare parts and support
- philosophy of maintenance management

One method is to use a ratio of repair person-hours per machine operating hour. Once the total machine hours are known, the total maintenance person-hour requirement can be determined.

Example 8.8:

One dozer requires 0.4 maintenance person-hours per operating hour (see Table 8.10). For a fleet of six dozers working 3,000 hours per year, the maintenance requirement is

$$0.4 \times 6 \times 3{,}000 = 7{,}200 \text{ maintenance person-hours per year}$$

A maintenance fitter is at work for 48 hours per week, but 8 hours of this time is lost on nonmaintenance tasks, such as training and safety. Thus, if a maintenance fitter works 40 hours per week, 45 weeks per year, the average time spent on maintenance tasks is 1,800 hours per year. In that case, four maintenance fitters would be required for the dozers.

The ratios are determined from handbooks, from historical records, or by back calculation from the maintenance repair labor cost per machine per operated hour. The ratio changes with the duty of the machine, so in a detailed study the ratios must be applied to individual operations, not just to the fleet as a whole.

Ratios include service labor (e.g., fitting the bucket teeth, resocketing ropes, etc.), since most of this service work can be undertaken by the maintenance or operating crews themselves during the normal course of their work. Table 8.10 includes estimates commonly used to relate maintenance person-hours to machine operating hours. Estimates in this table include *all* maintenance personnel, including mechanical fitters, electricians, auto-electricians, and instrument technicians.

EXAMPLE EQUIPMENT COST SCHEDULES

For the purposes of concluding this chapter, the costs and major operating parameters have been built up for a 16-m^3 hydraulic excavator working in a hypothetical mining environment. These schedules include components covering a whole range of equipment, even if some components are not relevant to the example given. (For example, electricity costs are shown, even though this excavator does not operate on electricity.)

Figure 8.3 shows a typical summary schedule. The ownership cost in this schedule is drawn from the sample buildup of costs in Figure 7.1 (p. 96), and the operating costs are drawn from Figure 8.4.

Figure 8.4 sets out the derivation of the operating cost of the equipment using the factors in Table 8.10 and the method of calculation described in the "Operating Cost Data" section of this chapter.

Figure 8.5 tabulates working hours and allowances for maintenance, standby time, and work delays.

Machine and Model:	Hydraulic Excavator—16 m^3
Date:	March 1998
Name of Company:	ABC Mining Company, Inc.
Project Name:	Typical Open Cut Mine
Application:	Front-Shovel Configuration Loading 170-t Rear Dump Trucks
Material Characteristics:	Well-Blasted Sandstone Density .... 2.4 t/m^3 (150 lb/ft^3) Swell .... 35 % Bucket Fill Factor .... 1.05 Operator Skill .... Average
Job Conditions:	Maintenance Support .... Good Ground Conditions .... Average Abrasiveness .... High Wear Fuel Usage .... High Consumption
Specifications:	Mass .... 290 t Rated Power (diesel powered) .... 1,120 kW Capacity .... 16 m^3
Cost Summary	
Ownership Costs: (see Figure 7.1)	Purchase Price .... $3,950,000 **Owning Cost .... $262.41 /operating hour**
Operating Costs: (see Figure 8.4)	Materials Only .... $230.25 /operating hour Maintenance Labor .... $50.00 /operating hour Operating Labor .... $55.00 /operating hour **Total Operating .... $335.25 /operating hour**
TOTAL	**Owning and Operating Cost $597.66 /operating hour**

FIGURE 8.3 **Typical equipment cost (summary) schedule**

Machine and Model:	Hydraulic Excavator—16 m^3
Source of Data/Reliability:	Derived From Formulas; Average Reliability
Power	Energy Unit Cost .... _____ /kW-h
	Average Consumption .... _____ kW/h
	Energy Cost .... _____ /hour §
	Demand Unit Cost .... _____ /kVA per month
	Demand.. _____ kVA per month
	Demand Cost .... _____ /hour §
Fuel	Fuel Unit Cost .... $0.25 /l
	Power .... 1,120 kW
	Load Factor... 0.5
	Average Consumption .... 168 l/h
	Fuel Cost .... $42.00 /hour §
Lube	Percent of Fuel Cost .... 25 %
	Lube Cost .... $10.50 /hour §
Tires	Cost per Set .... $_____
	Life per Set of Tires .... _____ hours
	Tire Cost... $_____ /hour §
Wear Items	Wear Cost (Allowance) .... $19.75 /hour §
Repair Parts	Expected Life .... 30,000 hours
	Repair Factor.............................0.25×10^{-4}
	Job Factor 1.0
	Repair Parts Cost .... $98.75 /hour §
Major Overhaul	Frequency .... 10,000 hours
	Percent of Capital Cost .... 15 %
	Overhaul Cost .... $59.25 /hour §
Maintenance Labor	Maintenance Ratio .... 1.5 hours/operating hour
	Maintenance Person-Hours Required.... 4,500 hours
	Annual Maintenance Cost .... $50,000 /person-year
	Person-Hours per Year .... 1,500 hours/year
	Hourly Labor Cost............................ $33.33 /hour
	Maintenance Labor Cost .... $50.00 /hour §
Operating Labor	Operators per Shift................................ 1.0 per shift
	Number of Shifts per Day 3.0
	Operator Ratio.. 1.0
	Cost per Year .... $55,000 /person-year
	Operating Hours of Equipment............. 3,000 hours/year
	Operator Cost .... $55.00 /hour §
Summary	**Operating Costs** (Items Denoted by §)
	Energy Cost .... $_____ /hour
	Demand Cost.................................. $_____ /hour
	Fuel Cost .... $42.00 /hour
	Lube Cost .... $10.50 /hour
	Tire Cost... $_____ /hour
	Wear Cost (Allowance) .... $19.75 /hour
	Repair Parts Cost .... $98.75 /hour
	Overhaul Cost .... $59.25 /hour
	Maintenance Labor Cost .... $50.00 /hour
	Operator Cost .... $55.00 /hour
All Operating Cost Items:	**Total Operating Cost .... $335.25 /hour**

FIGURE 8.4 Typical equipment cost schedule: operating cost data

Machine and Model	Hydraulic Excavator—16 m^3	
Calendar Days:	Total Days	365 days
Less Idle Time:	Weekends	104 days
	Mine Shutdown	0 days
	Public Holidays	11 days
	Scheduled Ordinary Time	250 days/year
Scheduled Time per Day:	Repair/Unscheduled Time	0 days
	Hours per Shift	8 hours
	Shifts per Day	3 per day
	Scheduled Time per Day	24 hours/day
Scheduled Hours:	Schedule Hours	6,000 hours
	Mechanical Availability	75 percent
Less Maintenance:	Planned Maintenance	750 hours/year
	Unplanned	750 hours/year
	Other	0 hours/year
	Subtotal Maintenance (Repair Hours)	1,500 hours/year
Available Hours:	Available Hours	4,500 hours/year
	(Equivalent to)	562.5 shifts/year
Less Standby:	Weather (approx. 5 days/year)	120 hours/year
	Industrial (approx. 15 days/year)	360 hours/year
	No Operator (Assume Nil)	_____ hours/year
	Conveyor Moves	_____ hours/year
	Not Required	_____ hours/year
	No Power	_____ hours/year
	Await Supporting Equipment	_____ hours/year
	Meal Break (40 minutes/shift)	375 hours/year
	Shift Change (20 minutes/shift)	188 hours/year
	Preshift Service (5 minutes/shift)	63 hours/year
	Subtotal Standby Time	1,106 hours/year
Operating Hours:	Operating Time	3,394 hours/year
	Utilization	75 %
Less Work Delay:	Fuel and Service (5 minutes/shift)	63 hours/year
	Dead Heading	_____ hours/year
	Positioning	_____ hours/year
	Minor Delays	331 hours/year
	Delay Hours	394 hours/year
Operating Hours:	Work Time	3,000 hours/year
	Work Delay	88 %

FIGURE 8.5 Typical equipment cost schedule: working hours and production data

CHAPTER 9

Investment Decisions

No mining company commits to a major investment without a thorough analysis to support the decision. Moreover, companies do not rely on only one measure of value to arrive at this decision. Once the technical work is finalized, the computational effort to generate a full suite of economic indicators is relatively minor. This chapter demonstrates how the discounted cash flow tools described in Chapter 5 are typically applied for investment decisions.

Along with financial analysis using DCF models, there are at least two other evaluation tools applied by investment strategists examining major mining projects. The first tool, referred to as the payback method, examines the cumulative flow of cash into and out of the project. Rather than focusing on the *return* from these cash flows, the payback method focuses on the *time* it takes until the initial outflows are recovered. Payback is examined later in this chapter (p. 132).

The second tool aims to provide information about the sensitivity of the project to changes in important data. It asks "what if" questions, such as: What would happen to the return on investment if the selling price changed? Sensitivity analyses are usually prepared from a base case discounted cash flow model, with changes to the input data translating into changes in net present value, internal rate of return, discounted average cost, or payback period. Sensitivity analysis is examined later in this chapter (p. 135).

All of these tools are forward-looking indicators to aid decision making. At the same time, published projections are also prepared by using accounting rules, since these rules will be the ones used to compare planned and actual performance once the project starts. These accounting rules sometimes present actual performance in ways that are contrary to expectations. Some of these issues are examined later in this chapter (see "Management Cost or Accounting Cost?" on p. 139).

DCF COMPARISON OF TWO ALTERNATIVES

A sample discounted cash flow tabulation for a mining project was set out in Table 5.2 (p. 54). The most important characteristics from this table are summarized in Table 9.1. In this tabulation, a hypothetical gold mine produces up to 50,000 oz per year and, at the expected selling price of $500/oz, yields a 15% return on the original $15 million investment.

Although this tabulation is quite typical, very few projects proceed without several alternatives being seriously considered. There are always various ways to exploit any deposit. A small company can maximize the use of contractors, minimizing the capital costs but at the expense of higher operating costs. If the amount of capital were not a problem, more capital-intensive alternatives could be chosen to provide reduced sensitivity to price changes.

This section examines the typical comparison of two such alternatives. The first case–case A–is represented by the sample discounted cash flow from Table 9.1. The second case–case B–is a more capital-intensive case. In this case, the same production is achieved by using a method that requires more capital. The extra $10 million of capital in this case yields operating costs about 20% less than in case A. Table 9.2 shows the discounted cash flow from case B in the same format as the original Table 5.2. Apart from the difference in the numbers, Table 9.2 is almost identical to Table 5.2. The data have again been deliberately chosen so that the project yields a 15% return at a gold price of $500/oz.

Table 9.2 has also been prepared to illustrate the case of tax losses. Note the tax treatment in the first 2 years of project life. In Table 9.2, case B incurs a loss in the first year because the higher depreciation charges exceed the

TABLE 9.1 Gold project: case A discounted cash flow (summary)

	Year					
	0	1	2	3	4	5
Production, *oz*		30,000	50,000	50,000	50,000	45,000
Operating revenue at $500/oz, *thousand $*		15,000	25,000	25,000	25,000	22,500
Operating expenses, *thousand $*		10,598	17,762	19,339	21,073	20,882
Operating profit, thousand $		4,402	7,238	5,661	3,927	1,618
Capital expenditure, *thousand $*	15,000					
Salvage value, *thousand $*						3,005
Income tax payable (at 35% tax rate), *thousand $*		97	1,486	1,223	824	167
Net cash flow, *thousand $*	(15,000)	4,305	5,751	4,439	3,103	4,455
Discounted cash flow (at 15% return on investment), *thousand $*	(15,000)	3,744	4,349	2,919	1,774	2,215
Net present value, $	**0**					

Note: All numbers in parentheses indicate negative values. Complete information is in Table 5.2 (p. 54).

TABLE 9.2 Gold project: case B discounted cash flow

	Year					
	0	1	2	3	4	5
Production, *oz*		30,000	50,000	50,000	50,000	45,000
Operating revenue at $500/oz, *thousand $*		15,000	25,000	25,000	25,000	22,500
Operating expenses, *thousand $*		9,109	14,737	15,693	16,720	16,241
Operating profit, thousand $		*5,891*	*10,263*	*9,307*	*8,280*	*6,259*
Capital expenditure, *thousand $*	25,000					
Tax depreciation this year at 27.5% declining balance, *thousand $*		6,875	4,984	3,614	2,620	1,899
End-of-year written-down value for tax purposes, *thousand $*		18,125	13,141	9,527	6,907	5,008
Salvage value, *thousand $*						5,008
Taxable profit for this year, *thousand $*		(984)	5,279	5,694	5,661	4,359
Assessed profit for tax payable, *thousand $*		0	4,294	5,694	5,661	4,359
Income tax payable (at 35% tax rate), *thousand $*		0	1,503	1,993	1,981	1,526
After-tax profit, *thousand $*		(984)	3,775	3,701	3,679	2,834
Net cash flow, *thousand $*	(25,000)	5,891	8,760	7,315	6,299	9,741
Discount factor (at 15% return on investment)	*1.0000*	*0.8696*	*0.7561*	*0.6575*	*0.5718*	*0.4972*
Discounted cash flow, *thousand $*	(25,000)	5,122	6,624	4,809	3,602	4,843
Net present value, $	**0**					

Note: All numbers in parentheses indicate negative values.

operating profit. No tax is payable. The loss reduces the taxable profit for the following year.*

In Tables 9.1 and 9.2, the net present value is shown as zero because the tabulation was set up to derive the internal rate of return. In both cases the internal rate of return is 15%.

A slightly different calculation is necessary to derive the NPV. Recall that the true (expected) NPV is the value added. The value added is the amount of money returned from the project above the cost of capital to fund the project. Thus, the true (expected) NPV is the cash flow discounted at the cost of capital.

In this case, assume the cost of capital is 8%. Table 9.3 shows the NPV for both cases calculated at this discount rate, along with other characteristics relevant to the decision. The question is, which is the preferred project?

* This method of carrying forward tax losses applies in most countries in the world. In some countries (e.g., the United States) tax losses can give rise to negative taxes—but only to the extent of taxes paid previously. Since new projects may not have past tax payments, they may not be able to take advantage of this tax treatment. The overall return for any project is enhanced if tax losses can be applied and taxes reduced in the year incurred rather than carried forward to future years. Because highly capital-intensive projects frequently show (tax) losses in the initial years, this is one reason many projects with multiple-company ownership are set up as joint ventures rather than stand-alone legal entities. In a joint venture, all cash flows are apportioned according to the rules of the venture, allowing tax losses to offset profits elsewhere within the same company.

TABLE 9.3 DCF comparison for cases A and B

Characteristic	Case A	Case B	Difference in Results
Production	Same for both cases		—
Operating cost as % of revenues	*78.2*	*63.6*	—
Initial capital cost, $	15,000,000	25,000,000	10,000,000
Internal rate of return, %	*15*	*15*	*15*
Net present value at 8% discount rate, $	2,753,000	5,031,000	2,278,000

With respect to return on investment (or internal rate of return), there is no difference between the projects. From a value added (or net present value) perspective, case B has the highest value—but since the two projects have different capital requirements, they cannot be compared on the basis of NPV alone. To make a choice, the difference in capital requirements must first be reconciled.

The easiest way to reconcile this difference is to undertake a small thought experiment. Assume first that the company "intends" to proceed with case A. Now think about case B. The choice to *instead* select case B is a choice to expend *an additional* $10 million of capital. This marginal capital will again yield an expected 15% return as shown and will add $2.278 million of value. These marginal values are also shown in Table 9.3. Whether case B is a better choice or not depends on how attractive this extra $10 million of investment is compared to other alternatives available.

Several questions must be asked. Where does this extra $10 million come from? If the company already has the funds from internal sources and does not apply them to case B, then what else will it do with the funds? If, on the other hand, the company does not already have the funds (and would have to raise them on the debt or equity markets), then what will this cost?

If the company has ample funds available from its own sources, and if the expenditure is small in comparison with the firm's size, then the answer is quite straightforward. Applying the extra $10 million to this project (selecting case B) is attractive if the NPV of this marginal investment ($2.278 million) exceeds the value added from the best alternative application of the funds.

If the expenditure is large in comparison with the size of the firm, however, then the answer is not so straightforward. Perhaps the first $15 million of capital can be raised at an 8% cost, but attempting to raise an additional $10 million might be problematic. Its effective cost might be more than 8%. In this case the *extra* NPV implied by the value in Table 9.3 is incorrect. A higher discount rate must be used for the higher-capital-cost case. For example, if the cost of capital for a $25 million investment were 11% rather than the 8% used for case A, then the NPV of case B would be less than that of case A. Case A would be the preferred method of mining from this economic perspective.

In comparisons of cases having different capital and operating cost alternatives, the conflict between internal rate of return and net present value occurs quite frequently. However, once the differing capital requirements are incorporated by using the logic just described, most conflict resolves. If the conflict cannot be resolved, then what choices are available? Two options are commonly followed:

1. Examine the other indicators described in the balance of this chapter.
2. Recognize that the apparent objectivity and quantitative nature of DCF tabulations are in fact based on many subjectively derived inputs, and adopt a pragmatic view.

Both options are recommended.

The pragmatic view recognizes that in the process of estimating the revenues, operating costs, future inflation, future taxation, and production capability, there are many unknown influences. If, after reconciliation of total capital requirements, the IRR indicator seems to suggest one choice and the NPV indicator seems to suggest a different choice, then almost certainly the difference will be minor. Slight changes in one or more of the inputs will change the result to favor one choice over the other by both indicators. This procedure is *not* recommended. The recommended procedure is simply to *be honest* and choose the alternative that is most liked by the people responsible for implementation. Engineers and finance managers always like to present results in strictly analytical ways, maintaining an illusion that objective, quantitative measures are all that are important. The pragmatic and honest approach is to recognize that easily understood and mechanically derived characteristics are only the first step. If these characteristics are insufficient to differentiate one project from another, then subjective measures (for example, that personnel simply *prefer* one alternative to the other) are probably more important than additional quantitative–and apparently objective–indicators pursued in even more intricate detail.

One further characteristic differentiating choice is shown in Table 9.3–the operating cost as a proportion of revenues. Many analysts use operating costs as a primary guide to compare projects. If two projects are each supplying the same market and selling their output at the same price, then the mine with the lowest operating cost generates more cash with each sale. This puts low-operating-cost projects at an advantage, particularly in the face of price declines.

Operating cost (or *cash* cost) is primarily a survivability indicator rather than an indicator of profitability. Consider an operating mine and what happens when the selling price declines. Most of the capital is sunk cost and is worth little in any other application. If the price declines, then the fixed depreciation write-offs quickly reduce the tax liability to nil. If there are no taxes (as outflows) and if the salvage value of equipment is near zero, then the only cash outflows are the operating cost. Thus, cash margin (the difference between selling price and operating cost) becomes a proxy for cash flow. Of two projects, in a survivability situation, the one with the lowest operating costs is relatively more valued.

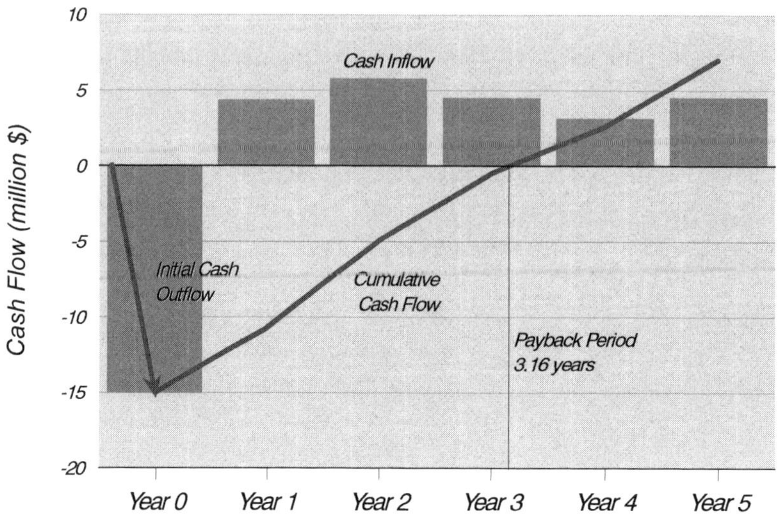

FIGURE 9.1 Cash flow and cumulative cash flow by year for case A

Is this a reliable measure? Cash operating margins are favorite comparative tools of stock analysts, but they need to be applied cautiously. An underground mine—where almost all the capital is in shafts and drives—has very little salvage value when the mine closes. Even temporary closures are problematic because of the ongoing cost of ventilation and drainage. An open pit mine—with owned equipment that can be readily sold—will not continue to operate where selling prices only barely cover operating cost. Moreover, care and maintenance (temporary closures) of open pit mines are often a realistic option because the ongoing support costs may not be substantial. Thus, the cash operating margin is a less reliable indicator of survivability in the case of open pit mines than it is for the survivability of underground mines.

PAYBACK

In addition to the criteria of net present value and internal rate of return, a supporting criterion—sometimes the only criterion—used for decision making is the payback period. The payback period is the time it takes a project to return to the investor the money that is put into the venture. The faster the payback, the less time that the owner's investment is at risk.

Calculation of the payback period is quite straightforward once a discounted cash flow has been prepared. The cash flows are simply plotted in cumulative form starting from zero expenditure before project commitment. Initial cash flows are invariably cash outflows—i.e., *negative* cash flows. The payback period is the time it takes for the cumulative cash flow to again become positive.

Figure 9.1 shows the year-by-year and cumulative cash flows from case A of the sample gold-mining project discussed in the previous section of this chapter.

TABLE 9.4 Cash flow tabulation for payback period

	Year					
	0	1	2	3	4	5
Case A:						
Net cash flow, *thousand $*	(15,000)	4,305	5,751	4,439	3,103	4,455
Cumulative cash flow, *thousand $*	(15,000)	(10,695)	(4,944)	(505)	2,598	7,053
Payback period: 3.16 years						
Case B:						
Net cash flow, *thousand $*	(25,000)	5,891	8,760	7,315	6,299	9,741
Cumulative cash flow, *thousand $*	(25,000)	(19,109)	(10,349)	(3,035)	3,265	13,005
Payback period: 3.48 years						

Note: All numbers in parentheses indicate negative values.

For projects with one initial capital outlay at the start followed only by positive cash flows, the payback is an excellent way to visually portray the flow of funds. Table 9.4 shows the cash flow and cumulative cash flow for both cases A and B from the preceding section of this chapter. If payback period is the only selection criterion, how does the less capital-intensive case (case A) compare with the more capital-intensive case (case B)? Case A has a payback period of 3.16 years, whereas case B has a payback period of 3.48 years. Thus, in this instance case A is favored over case B.

The payback calculation yields more information than just the payback period, however, and the method should not be viewed only as a mechanism to calculate this period. The value of the payback method is evident in the way the method illustrates the flow of funds. For instance, Figure 9.2 again shows the two cases with their cash flow profiles. Consider first the shape and slope of the two lines in Figure 9.2. The slope of the line represents the strength of the cash flow—the rate at which cash is flowing into or out of a project. For cash inflows from year 1 onward, it indicates the cash margin between selling price and cost of production (including taxes). On this measure, case B is favored over case A—particularly if there is potential for the mine to continue after year 5. Mines that are *planned* for some set time but have the potential for a life beyond the initially planned period benefit from more capital-intensive processes.

Another factor used for decision making is the maximum negative cash flow, which represents the maximum amount of funds that must be sourced external to the project. In this case the maximum negative cash flow is the initial capital, but normally mining projects start to generate cash before all the expenditures on mining plant have been made. Mines that start shallow and progress deeper fit this classic cash flow profile, where subsequent investment is partially or fully funded from initial or retained earnings.

Most investment strategists focus on the maximum negative cash flow rather than the total capital. It is the maximum negative cash flow that determines the funding requirement and the amount of money being put at risk. Thus, a

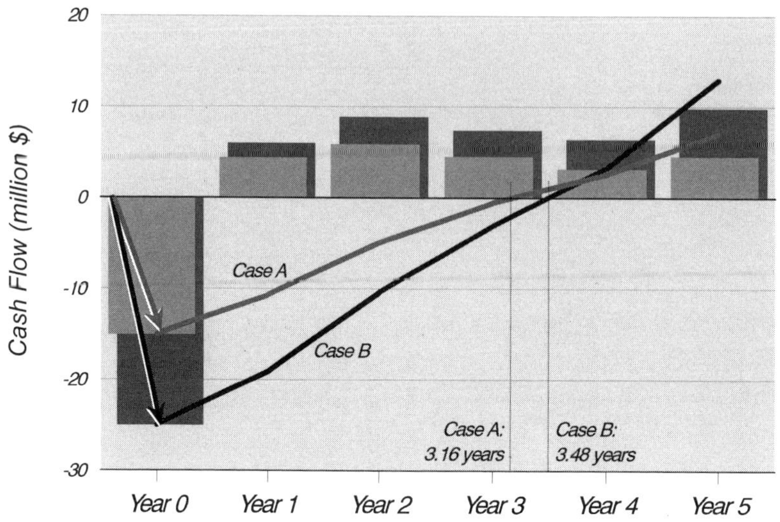

FIGURE 9.2 Payback comparison of cash flow profiles

project that can be started at a small production rate and then continually expanded can use the cash flow from earlier years to fund the expansions. This limits the amount of external funding necessary, but because the initial capital may not be returned for many years it results in an apparently poor payback period.

Consider, for example, a project like case A that can be expanded to three times its size. Two options are available. In the first case, the three-times-higher production rate can be planned from the start, requiring three times the initial capital. Alternatively, the project can be started as for case A, then expanded again in year 2 (higher production from year 3 onward), and then expanded again in year 4 (higher production from year 5 onward). For simplicity, assume that the cash flows in year 4 of case A from Table 9.4 continue indefinitely after year 4. The cash flow profile for both of these cases is shown in Figure 9.3.

On a superficial examination, the case of expanding in three increments looks less attractive, with payback taking 72% more time. However, this result overlooks two important advantages applying to the incremental expansion case:

1. If the funding for the expansion is sourced from cash flow, the maximum negative cash flow is reduced to only 44% of the alternative case. Financing costs will be less. The risk is lower.
2. The decision to expand is predicated on success in the first increment in production. The decision (to expand again) in years 2 and 4 is an option that can be rescinded costlessly. Indeed, since performance criteria from the first stage of production can be used to plan the subsequent stages of production, then these subsequent stages are likely to be more efficient than they would be if committed to from the start.

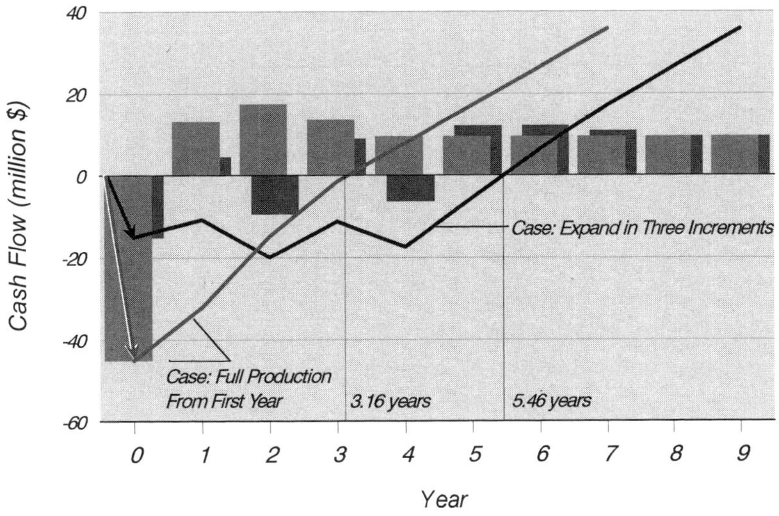

FIGURE 9.3 Cash flow profiles of expanded projects

Payback profiles are valuable adjuncts to the other techniques of investment evaluation, but payback period on its own may not faithfully indicate genuine investment objectives of risk reduction and options to change.

Two criticisms of the payback method, if it is to be used in any quantitative way, are (1) that it does not account for the time value of money and (2) that it does not account for the cash flows occurring after the payback period. These are valid criticisms. To overcome the first of these problems, some investment analysts discount the cash flows prior to calculating the payback period. If this discount rate is the interest rate that could be earned by putting the funds in the bank, then the result of the calculation is the time until the return exceeds the return from bank-invested funds.

All of these ideas have merit. However, in practice, return-on-investment criteria are faithfully addressed in the standard DCF calculation; provided both techniques (i.e., DCF techniques and the nondiscounted payback techniques) are used, applying discounting into the payback calculation probably adds little to the overall analysis.

SENSITIVITY ANALYSIS

Before a commitment is made for a major investment, decision makers must address the "what if" questions. For instance, if the selling price declines by $X, what will that do to the return on investment?

These types of questions are usually addressed with a sensitivity analysis. Sensitivity analyses look at varying one or more of the input variables to determine how much the return on investment (or some other decision criterion) changes. Figure 9.4 shows a simple sensitivity analysis using the cash

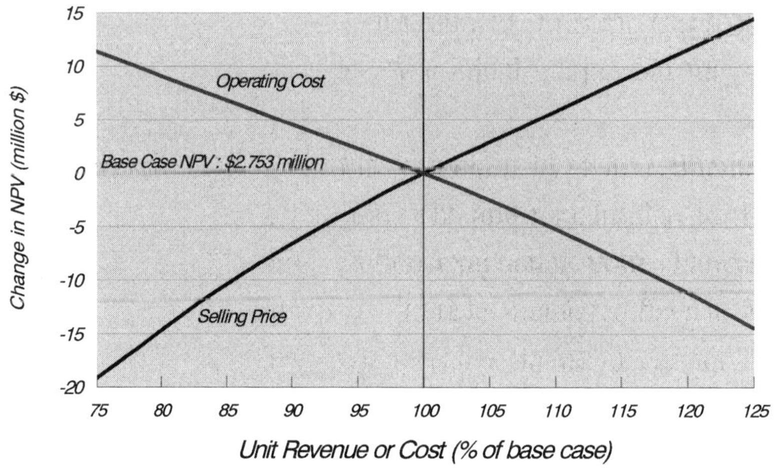

FIGURE 9.4 Sensitivity analysis: net present value

flow data from case A described previously. In this figure, if the operating costs turn out to be 10% greater than planned and nothing else changes, then the NPV will decline by approximately $5.2 million. If the selling price turns out to be 10% greater than planned and nothing else changes, then the NPV will increase by approximately $5.7 million. Similar diagrams can be prepared showing the change in return on investment.

Two difficulties are evident:

1. The analysis on its own does not provide any guidance for the likelihood of these events happening. In this case, is a 10% increase in costs equally as likely to occur as an increase in revenues? Projects may be very sensitive to certain assumptions, but if there is little likelihood of the assumption turning out to be wrong, then this is irrelevant.
2. Sensitivity analyses invoke the ceteris paribus assumption, i.e., that all other things remain unchanged. This is the biggest difficulty, since mines are changing all of the time in response to changes in the external environment. If the price of fuel oil rises, the mine will preferentially use more electrically powered equipment. If the selling price rises, lower-grade ores will be incorporated into the mine plan, which may then be changed dramatically. All other things *do not* remain unchanged.

For these reasons, the simple sensitivity study is only of limited use. Nevertheless, sensitivity analysis does provide a useful starting point for two extensions of its application: probabilistic analysis and relative sensitivity analysis.

The probabilistic analysis addresses the probability question. If the *probability* of some change (in an input characteristic) can be obtained or can be estimated, then this analysis allows a probability distribution of the NPV or IRR to be drawn. An example of this is set out in Chapter 14. Probabilistic sensitivity analysis has found only limited application in the mining industry to date, principally because of the difficulty in obtaining the characteristics of the input variables. In addition, the complexity of the analysis, combined with the

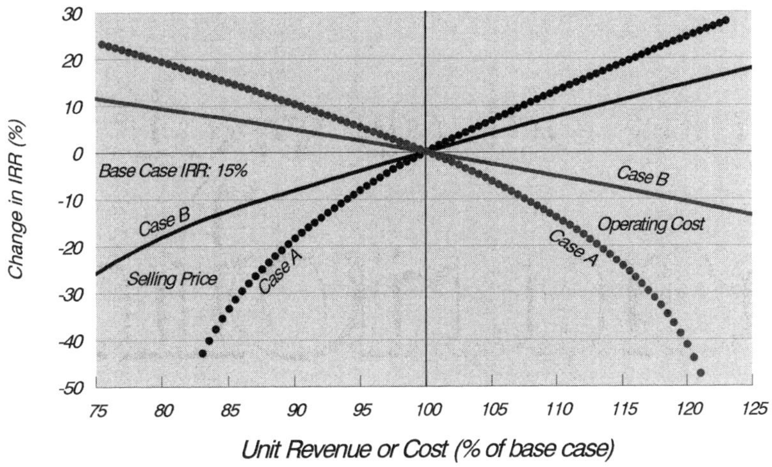

FIGURE 9.5 Relative sensitivity analysis: IRR indicator

difficulty in interpretation of results, makes decision making based on this type of analysis somewhat problematic.

Sensitivity analysis can be used quite productively to highlight the *relative* differences between project alternatives. In this context the sensitivity of the project to changes in inputs becomes the decision criterion.

To understand this application, consider again the shortcoming highlighted earlier—the ceteris paribus assumption. This was labeled a shortcoming because not all other things remain unchanged. The question can be rephrased as follows: If circumstances allow only a very limited ability to change, then is this changeability or lack of ability to change a point of differentiation between projects? If projects cannot be designed to allow for change, choices can at least be made favoring projects that are less susceptible to unexpected change. Mines operating in remote regions or mines subject to restrictive industrial or finance agreements frequently have such limitations. Mines that have been financed by using nonrecourse funds (whereby the lenders can be repaid only out of the project's own cash flows) are very susceptible to such restrictions because the lenders must typically approve any changes to the mine plan before implementation.

Figure 9.5 shows the change in internal rate of return for cases A and B, demonstrating that case B is less sensitive than case A both for changes in selling price and for changes in operating cost.

Sensitivity analysis can also be undertaken on the payback period. Figure 9.6 shows the same cases, demonstrating that case B is less sensitive than case A.

If mine management has limited scope to change, this might influence the choice of one case or another. An alternative approach with sensitivity analysis is to examine the opportunity to offset a change in one of the uncontrollable inputs with changes elsewhere that *are* under management control. For instance, two common inputs that management cannot control are the selling

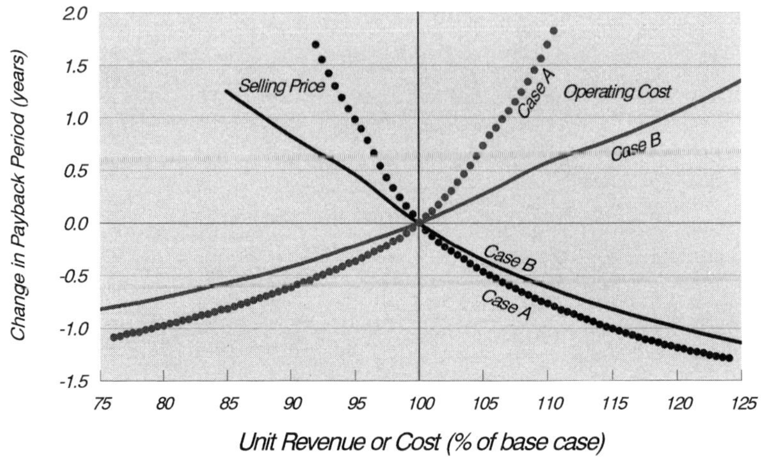

FIGURE 9.6 Relative sensitivity analysis: payback indicator

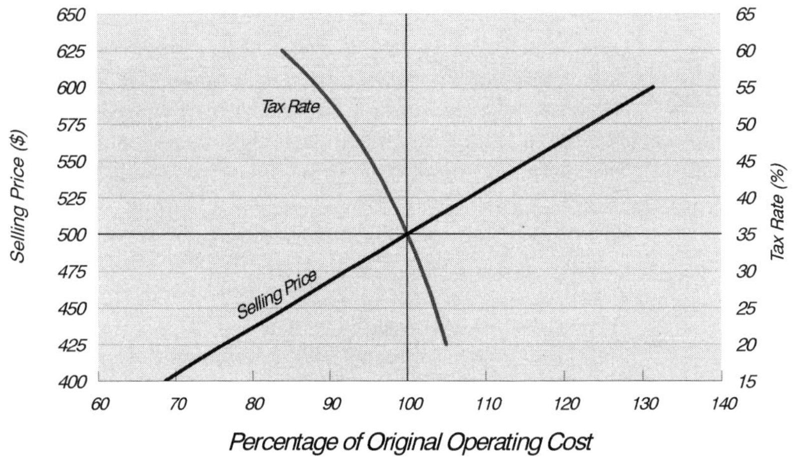

FIGURE 9.7 Maintaining investment return with change

price of the product and the tax rate applied. This variation on the sensitivity would ask the following question: If the tax rate (or any other uncontrollable factor) changed, what must the company do to mine operating costs (or any other factor that *can* be controlled) to maintain the expected return on investment?

Figure 9.7 shows the required change in operating cost for case A corresponding to changes in tax rates or selling price to maintain the expected return on investment. In the figure, if the tax rate increases from 35% to 45%, operating costs would need to drop to 95.3% of their original value (a reduction of 4.7%) to maintain the same level of profitability. A change in selling price of $14.90 corresponds to a change in operating cost of the same order as this change in tax rates. Governments sometimes use these relationships when

TABLE 9.5 DCF analysis of a reclamation project

	Year				
	0	1	2	3	4
Production		4,968,809	4,907,008	4,845,977	4,785,704
Revenue **at $0.1871 per unit of production**, $		929,792	918,228	906,807	895,529
Capital expenditure, $	750,000				
Salvage value, $					75,000
Book value, start of year, $		750,000	581,250	412,500	243,750
Book value, end of year, $		581,250	412,500	243,750	75,000
Depreciation, straight line, $		168,750	168,750	168,750	168,750
Total operating costs, $		*602,678*	*618,576*	*635,225*	*652,663*
Operating profit, $		327,114	299,652	271,582	242,865
Profit for tax purposes, $		158,364	130,902	102,832	74,115
Tax payable, $		55,428	45,816	35,991	25,940
Net after-tax operating profit, $		102,937	85,086	66,841	48,175
Cash flow, $	(750,000)	271,687	253,836	235,591	291,925

Note: Numbers in parentheses indicate negative values.

adjusting taxes and tariffs (i.e., when implementing microeconomic reform). The objective in this microeconomic reform is to simplify the tax structure in a way that firms are indifferent to the change.

MANAGEMENT COST OR ACCOUNTING COST?

In most large organizations, all the information necessary for derivation of the discounted cash flow is available from data already collected for accounting purposes. However, if this information is to be used directly for management decision making, then the implications of some of the results need to be clearly understood. Accounting anomalies may make a poor manager look good and a good manager look poor—a situation that is inconsistent with holding managers accountable for cost elements under their control.

The difficulty is best illustrated with an example. Table 9.5 shows a summarized discounted cash flow tabulation of a dozer used for a reclamation project, similar to the discounted average cost calculation in Chapter 5 (see Table 5.6, p. 68).

Assume for a start that this project (which actually consists of only a single dozer undertaking reclamation) is commenced and through its entire life performs exactly according to plan. Table 9.6 shows the standard accounting results that would flow from such a situation. At the start of the project, the capital valuation of the dozer (i.e., $750,000) equated to the present value of the expected future cash flows. In other words, the investment was yielding a 15% return.

TABLE 9.6 Accounting results for standard DCF analysis

	Year			
	1	2	3	4
Accounting after-tax return (profit), $	102,937	85,086	66,841	48,175
Accounting after-tax return on assets, %	*13.72*	*14.64*	*16.20*	*19.76*
Accounting after-tax cost per unit, $	0.166	0.170	0.173	0.177
Accounting after-tax profit per unit, $	0.021	0.017	0.014	0.010
Accounting after-tax profit per unit, %	*11.07*	*9.27*	*7.37*	*5.38*

Now consider the situation at the end of the first year. Even though everything is according to plan, the return on assets is just 13.72%! It seems like management has failed to perform. Similarly, for years 3 and 4, the return on assets exceeds the target 15%. It seems like management is doing a great job when in fact they are working exactly to plan. The anomaly comes from the method of depreciation used.

From an economic perspective, the dozer should be valued at the end of year 1 according to the work or future value the company expects to get from it in its remaining life. The present value of the future cash flows at the end of year 1 (expressed in year 1 valuation terms) is $590,813, whereas the written-down value in Table 9.5 is $581,250. The straight line depreciation method is *understating* the "true" profit by $9,563 and calling it depreciation. As a result, the accounting profit shown on the first two lines of Table 9.6 *understates* the profit in the first 2 years and then *overstates* the profit for the remaining 2 years.

Indeed, if this convention is used, *even for projects that perform to expectations,* the after-tax profit as a percentage of sales *declines* throughout the machine life, while the after-tax return on assets employed *improves* throughout the machine life. The accounting profession is certainly aware of this anomaly (e.g., see Brealey and Myers [2003, p. 326]), but even if generally accepted accounting procedures may be able to overlook it, management decision making should not. Management guidelines based on the accounting definition of per-unit profitability bias business decisions in favor of newer equipment. Management guidelines based on the accounting definition of return on assets bias decisions in favor of older equipment.

For operational decision making, the criteria for asset valuation must be market based, and internal prices and depreciation schedules must be calculated accordingly. When the same opportunity cost of capital and all costs according to the projected investment plan are used, depreciation throughout the life of the asset should result in written-down values that each year balance the present value of the expected future cash flows. A company could "sell" its own equipment to itself at any time, use that equipment for the intended purpose, and find that—at that "purchase price" (internal asset valuation)—the equipment would yield the company's required return.

TABLE 9.7 Discounted average cost calculation with constant return on assets

	Year				
	0	1	2	3	4
Production		4,968,809	4,907,008	4,845,977	4,785,704
Gross revenue **at $0.1873/unit**, $		930,708	919,132	907,700	896,411
Capital expenditure, $	750,000				
Salvage value, $					75,000
Book value, start of year, $		750,000	595,047	431,809	258,982
Book value, end of year, $		595,047	431,809	258,982	75,000
Depreciation, $		154,953	163,238	172,827	183,982
Total operating costs, $		*602,678*	*618,576*	*635,225*	*652,663*
Unit operating costs, $		*0.121*	*0.126*	*0.131*	*0.136*
Operating profit, $		328,030	300,556	272,475	243,747
Profit for tax purposes, $		173,077	137,319	99,648	59,765
Tax payable, $		60,577	48,061	34,877	20,918
Cash flow, $	(750,000)	267,453	252,495	237,598	297,830
Present value of cash flow (at 15% IRR), $	(750,000)	232,568	190,922	156,225	170,285
Net present value, $	0				

Note: Numbers in parentheses indicate negative values.

Table 9.7 sets out a cash flow similar to Table 9.5 using an *economic* depreciation schedule, with accounting data similar to Table 9.6 again following (in Table 9.8).

Table 9.8 is not the same as Table 9.6 just with a new depreciation schedule. The changed depreciation rate also affects the tax payable and ultimately the unit cost of production to balance the cash flow.* In this table, the depreciation schedule was iteratively determined concurrently with the discounted average cost (unit revenue) calculation. With a depreciation schedule yielding a constant return on assets, the distorting effects of non-market-based valuations are removed, and the bias favoring older equipment is neutralized. The second line of accounting data (in Table 9.8) shows a constant 15% return on assets consistent with the original 15% return on the original investment.

The incorporation of market-based depreciation in operational decision making is not just a subtlety of academic interest—it forces a degree of alertness on

* In most jurisdictions there is no requirement for tax-based depreciation to be the same as depreciation for corporate finance purposes. Ordinarily a tax-based depreciation schedule will be adopted that minimizes *taxable* profit early to reduce the tax payable and improve early cash flow. Because of widely varying tax treatment of these issues around the world, the approach adopted in this example was to keep tax-based depreciation consistent with the *economic* valuation of partially worn-out equipment. As a result, the notional cost of production used in the example has increased by 0.1% over the case set out in Table 9.5. In practice, *economic* depreciation would *not* be used for tax purposes, and the calculated cost of production would be unchanged.

TABLE 9.8 Accounting indices for constant return DCF analysis

	Year			
	1	2	3	4
Accounting after-tax return (profit), $	112,500	89,257	64,771	38,847
Accounting after-tax return on assets, %	*15.00*	*15.00*	*15.00*	*15.00*
Accounting after-tax cost per unit, *$/unit*	0.165	0.169	0.174	0.179
Accounting after-tax profit per unit, *$/unit*	0.023	0.018	0.013	0.008
Accounting after-tax profit per unit, %	*12.00*	*9.71*	*7.14*	*4.33*

operational personnel that is missing when simple accounting measures are used. This alertness is vital to the management of change.

Compare, for instance, the similar, activity-based costing example from the Chapter 5 section entitled "Discounted Average Cost" (p. 64) to the preceding example with and without market-based asset management. Without market-based asset valuation, there is no incentive to use or even to dispose of older equipment if, by circumstance, the mine has too much equipment or inappropriate equipment. Such a circumstance is common in many industries subject to changing technology and varying cyclical and product quality demands. Equipment in the middle of its technical life is left unused—and *unplanned* to be used—but not written off because of reluctance to acknowledge capital write-downs. This reluctance is understandable—markets do not like unexpected charges against earnings—but is counterproductive if it results in continued "use" of economically unproductive assets. The return on productive assets has to cover the dead weight load of the unproductive assets.

Asset management for operational decision making has to value each item of equipment annually according to the expected return from use (or, if higher, from disposal).

CHAPTER 10 Operating Mine Case Study

An economic analysis for a new mine is mandatory. No major mining company approves the funding for new developments without a thorough understanding of the expected costs and return on investment.

For operating mines, the same rigor that applies to new projects is often missing. In many ways, economic analysis is more difficult and less useful—if a company's main product load-out facility just had a major failure, the company personnel do not need a big study to tell them that it is economical to replace it. Many site personnel also avoid detailed economic analysis out of concern that the analysis may not be supportive of their objective.

Another difficulty in applying discounted cash flow analysis to operating mines is that many decisions at these mines do not involve any change in mine output. Without change in mine output, there is no change in mine revenue and apparently (but incorrectly) no basis for discounted cash flow analysis. This perception is incorrect. In practice many site problems are more urgent and have a more economical solution than new capital expenditure at new mines. Without a thorough economic analysis, there is a risk of underdesign. Past underdesign due to the lack of thorough economic analysis is itself a contributor to the lack of confidence in future evaluations.

The case study in this chapter details a step-by-step set of interrelated decisions in an operating mine. Each decision is supported by an economic analysis, considering one or more aspects covered in preceding chapters. The case study is not complete—even simple real-life studies are more complex than can be portrayed in just a few pages of text.

The case study also introduces a number of strategic planning concepts that are examined in greater detail from Chapter 12 onward.

TECHNICAL ANALYSIS: WASTE REMOVAL

The case study involves a mine expansion, with additional waste removal of about 3 million m^3/year. (In this chapter, all waste volumes are expressed as bank, or unswelled, quantities.) The additional capacity is needed for the foreseeable future (the next 6 years or longer). The company has a proposal from a well-respected contractor offering to move any amount of waste at \$2.50/m^3. (This input to the calculation provides an initial opportunity cost—an essential ingredient in any economic analysis. Without this initial opportunity cost, the relevant "cost" is the value of the mine under the "do nothing" scenario.) Senior management will consider this proposal fairly, although to date they have always preferred to use their own equipment for regular long-term earthmoving and use contractors only for special projects or for peak loads.

The technical study of the earthmoving suggests using a large front-end loader coupled with 136-t-capacity rear dump trucks. The operating and initial capital costs of the proposed loader and truck are set out in Table 10.1. Operating costs for this equipment have been reconciled and shown consistent with the operating costs currently incurred while running the company's existing smaller loader and trucks.

No one fixed number of trucks matches the mine's requirement exactly—however, the mine schedules do not necessarily demand an exact match. If an optimum fleet turns out to have a production slightly more or slightly less than the nominal 3 million m^3/year, other equipment can be scheduled to make up the differences. For the typical haul cycles proposed, the estimated annual production from various sized fleets is set out in Table 10.2.

Since the total expenditure in this example will exceed \$10 million, a thorough discounted cash flow analysis of the proposition is necessary. Different equipment lasts for different amounts of time depending on type and schedule of use. This sort of evaluation is best undertaken by assuming a certain project life and adjusting end-of-project values. For the purposes of unbiased economic evaluation, equipment that is still worth something at the end of the project life is assumed to be sold at the book value, or written-down value, at the end of this time. Guidelines for this analysis are set out in Table 10.3.

The evaluation is undertaken using all capital costs and operating costs because, before purchase, *all* of these costs are variable. Once equipment is already in place, the "purchase price" of equipment being used in these calculations may be replaced by the opportunity costs of retaining the equipment. One such case is taken up in the "Asset Management Considerations" section later in this chapter.

The front-end loader matched with only three trucks does not produce an annual quantity consistent with mine requirements, but the four-, five-, and six-truck matches are all viable options. Figure 10.1 sets out the production from these various sizes of fleets in graphical form. The discounted average cost of production must be calculated for each of the options. This calculation, assuming the loader and trucks are the only equipment, has been

TABLE 10.1 Capital and operating costs for 136-t truck and front-end loader

Cost Item	136-t Truck	Front-End Loader
Initial capital cost, $	1,400,000	3,200,000
Expected life, *years*	8	6
Annual usage, *operating hours*	3,329	4,165
Operating cost per hour		
Operating labor, $	48.24	58.43
Maintenance labor, $	21.44	30.60
Fuel, $	25.08	52.03
Lube, oil, greases, $	3.76	7.81
Tire wear, replacement, $	18.55	15.91
Wear items, $	7.00	16.00
Repair parts, $	39.20	136.00
Major overhauls, $	17.50	48.00
Total operating cost, $/*operating hour*	**180.77**	**364.78**

TABLE 10.2 Loader/truck production estimates

Number of Trucks (136-t capacity)	Annual Production (million m^3)*
3	2.238
4	2.812
5	3.124
6	3.254

* Production rates are calculated by computer simulation. Clearly these rates are not known with a reliability appropriate for four significant figures of accuracy. Nevertheless, there is a good reason for *not* rounding off calculated numbers until the final presentation. Since many economic decisions are based on the marginal return, the *change* in production or cost may well be reliable to a higher order of accuracy than the actual average cost or average production. Rounding off numbers that have been calculated in a consistent fashion risks distorting these marginal calculations. In this example, numbers have been retained in full precision so that graphical presentation of marginal production and cost results will yield the smooth curves that are intuitively expected.

TABLE 10.3 Discounted cash flow guidelines for waste removal example

Cash Flow Feature	Comment or Value
Required return on investment	15%
"Project" life	6 years
Depreciation	Straight line over technical life of equipment
Salvage values	Assume recovery at written-down value of equipment
Corporate tax rate	39%

undertaken identically to the example in Chapter 5 (see "A Sample Discounted Average Cost Calculation," p. 66). All discounted average cost calculations were undertaken along with productivity calculations within TALPAC® (Runge Mining, Inc., Denver, Colo.), a computer program originally developed by the author for economic analyses of truck and loader productivity. These calculations are an integral part of the program. Table 10.4 sets out the results of this calculation, with average costs plotted against truck fleet size presented in Figure 10.2.

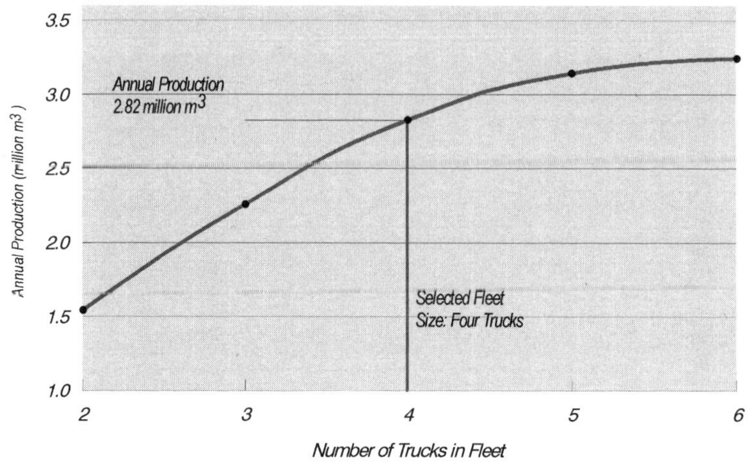

FIGURE 10.1 Annual fleet production

TABLE 10.4 Average costs of production for front-end loader and various numbers of trucks

Number of Trucks (136-t capacity)	Annual Production (m^3)	Costs of Production ($/m³)		
		Equivalent Capital Cost	Operating Cost	Total Cost
3	2,237,577	1.033	1.487	2.520
4	2,811,959	0.981	1.395	2.376
5	3,124,153	1.023	1.449	2.472
6	3,254,585	1.111	1.577	2.688

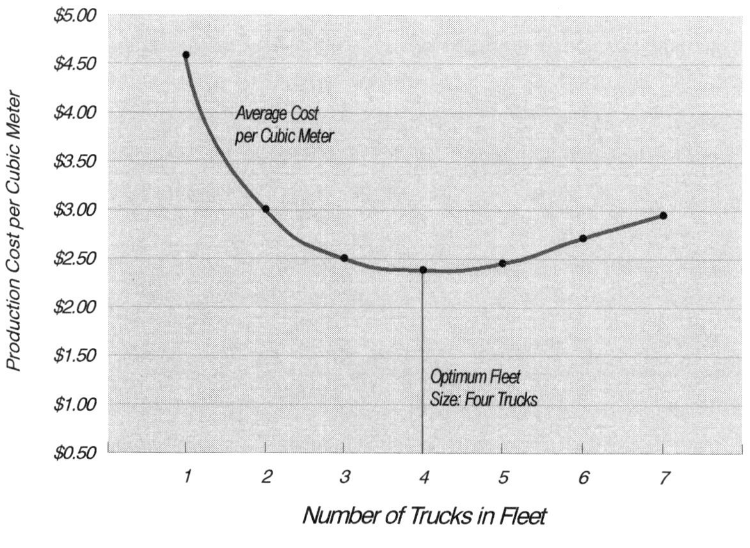

FIGURE 10.2 Fleet average costs of production

TABLE 10.5 Marginal costs with increases in fleet size

	Number of Trucks	Annual Production (m^3)	Marginal Production (m^3)	Average Cost ($\$/m^3$)	Total Cost Over 1 year ($)	Marginal Cost ($\$/m^3$)
	3	2,237,577		2.520	5,638,700	
Marginal case: 3 to 4 trucks			574,382		1,042,500	1.815
	4	2,811,959		2.376	6,681,200	
Marginal case: 4 to 5 trucks			312,194		1,041,700	3.337
	5	3,124,153		2.472	7,722,900	
Marginal case: 5 to 6 trucks			130,432		1,025,400	7.862
	6	3,254,585		2.688	8,748,300	

Figure 10.2 shows that the lowest cost of earthmoving in this application is for a fleet of four trucks matched with the loader. The five-truck match is also competitive. This suboptimum fleet has 11% higher production than the optimum-sized fleet, has only 4% higher cost on average, and is still of lower cost than using contractors.

A superficial assessment could easily recommend a fleet of four trucks, for which annual production is slightly less than required. Alternatively, a fleet of five trucks is viable, with production slightly more than required. Both cases move material at average costs lower than the cheapest other alternative (contractor).

The difficulty with this conclusion is that it focuses on *average* costs to the exclusion of marginal costs. The four-truck fleet—the base case—is indeed the alternative with the lowest (total) cost. Alternatives to the base case must be assessed on their marginal benefit and marginal cost, not their average cost. When the number of trucks in the fleet is less than the optimum, adding another truck lowers the average cost. The extra waste moved by the extra truck is lower cost than the waste already being moved. The reverse is true for increases to the truck fleet above the optimum. Table 10.5 sets out the calculation of this marginal production and marginal cost.

The addition of one truck to the fleet adds a constant annual cost but a declining increase in production. (The marginal annual cost shown in Table 10.5 should be the same with the addition of each truck. The actual figures shown vary slightly as a result of rounding.)

Figure 10.3 highlights this change in cost much more dramatically than Figure 10.2 by plotting marginal cost and average cost against annual production on the horizontal axis rather than showing the number of trucks on this axis. The marginal and average production costs associated with each fleet size are shown in the figure, connected by a smooth curve. Clearly the curves are simplifications since integral numbers of trucks have to be purchased. The only meaningful values on these curves are the ones that correspond to

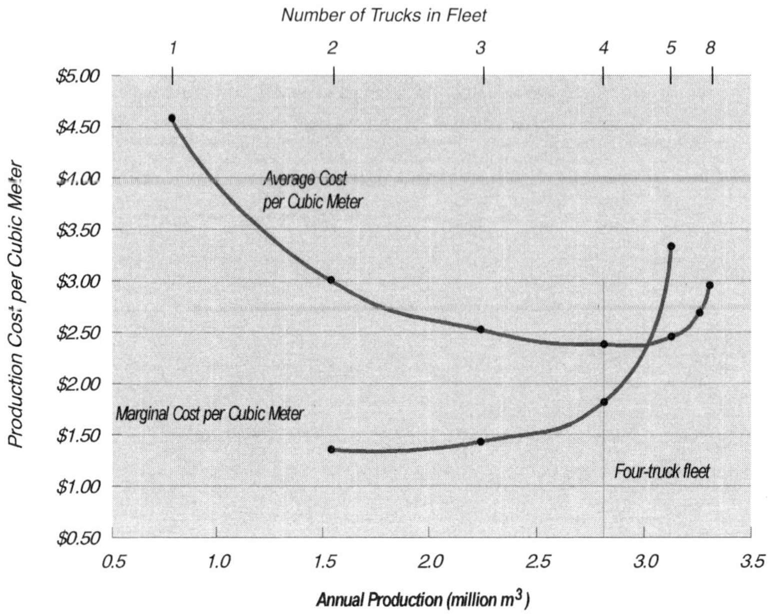

FIGURE 10.3 Per-unit average and marginal costs relative to annual fleet production

these integral numbers. The section "Economics of Operational Decisions" later in this chapter examines this marginal cost when equipment is already owned and where production rates *not* corresponding to fully utilized integral numbers of trucks are required. For the first part of this chapter, though, the use of curved lines linking discrete points on the graph simplifies the analysis and does not lead to erroneous results.

Although the average cost of the five-truck fleet is still less than contractor costs, the marginal cost of the waste moved by the fifth truck is substantially higher than that of having contractors move the same waste.

Recommendation 1: A four-truck fleet may be purchased, with an expected output of approximately 2.8 million m^3/year. If additional production is required, contractors should be used.

This recommendation could not have been deduced from accounting records even if the fleet were actually operating and monitored. The fifth truck is not a singularly expensive earthmover. When the fifth truck is added to the fleet, all five trucks move waste at the same cost per cubic meter, but the cost per cubic meter is higher than with four trucks. The marginal cost upon which the decision is made is the *change in total cost* occasioned by the addition of the truck to the fleet. These marginal costs are unobservable even by a perfect monitoring and accounting system. They have to be deduced.

STRATEGIC CHOICE

The analysis set out in the preceding section typifies most analysis in mines. It could be described as a technical study, with cost applied. If more sophisticated analysis is called for, then this is commonly undertaken by financial specialists remote from the mine site. Such analysis might look at changing depreciation rates or changed financing structures associated with different equipment tenders. The assumption in this approach is that changes to financial criteria are primarily refinements to a basically unchanging technical plan.

This assumption must be challenged. What were the guidelines for the initial evaluation? Did the results support these guidelines? If the guidelines were changed slightly, to what extent would the results change, if at all? If indeed the guideline of 3 million m^3/year were immutable, then perhaps no further analysis would be called for. This is not the case here, though, and is seldom the case in mining worldwide. Often initial guidelines are very ill defined and yet remain unquestioned at the start and throughout the entire evaluation.

The strategic planning process takes cognizance of a changing world. While recognizing the uncertainties inherent in this world, the process aims to develop schemes that have a greater adaptability and hence higher chance of achieving expectations in the face of change. Would this make a difference to the recommendation in this case? To address this sort of question, the narrowly defined case already looked at has to be placed in a broader context—a context that covers a wider range of cases for the most likely variables impacting mine economics. In most mining (and most any industry), the most likely changes in an uncertain future relate to output and price.

Throughout their life, almost all mines *expand* rather than contract. There is a good reason for this. Mining decisions involve a lot of uncertainty, and resolving this uncertainty at the start of the mine is very expensive or impossible. On the other hand, a mine that is economic to start at some small rate will, once it is in production, have a reduced cost associated with resolving these uncertainties. Drilling out an underground orebody from an adjacent crosscut is far cheaper than exploration from surface drillholes. Mines are usually started at suboptimum overall production rates for this reason alone.

The second reason mines usually expand (applicable in *most* areas of the world) has to do with capital structure. Fixed infrastructure associated with many mines located in remote regions can often service much higher production at little or no extra cost. If the mine is to commence at all, it has to commence at production rates substantially below the superficial optimum rate for the deposit. Except in rare cases, the economic forces in this environment are pushing toward expanded production. Perhaps an increase in waste output by 3 million m^3/year is all that can be envisaged now; however, if some alternative plan were almost as viable but allowed even greater expansion, then perhaps that plan would be a better option.

For completeness, Figure 10.4 shows the same haulage situation as used in Figure 10.3, with costs of production shown for a complete range of available

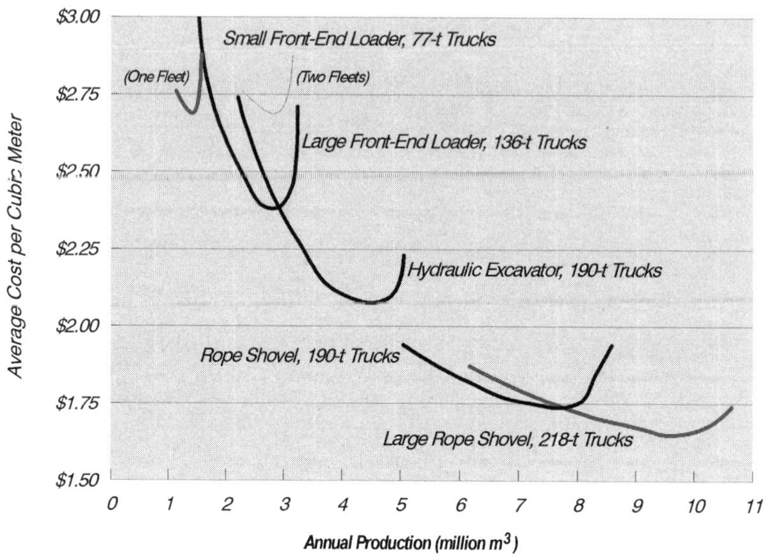

FIGURE 10.4 Average production costs for complete range of equipment

mining equipment. The lowest-production case has the highest costs and, as expected, uses the smallest equipment. As equipment sizes increase, cost-effectiveness improves. Although there are a number of reasons for this, the most evident is the savings in terms of labor cost. One truck driver in, for example, a 218-t truck costs little more in wages than a truck driver in a small, 77-t truck.

ECONOMICS OF CAPITAL UTILIZATION

Figure 10.4 immediately suggests a number of questions. The first of these is: At the annual production of 3 million m^3/year, should an "optimum" fleet of the 136-t trucks be selected when a suboptimum fleet of larger-size trucks may move the waste for a similar price? Since the mine is more likely to expand in the future rather than contract, what premium can the company afford for expandability? If the (narrowly defined) optimum fleet of 136-t trucks is purchased, what options are available if there is further expansion?

These questions are part of every economic evaluation. Very few real-world issues have outcomes that fall into some neat optimum category. All the potential solutions are suboptimum in some way, and recognizing the possibilities is one of the challenges to this sort of evaluation. The trade-offs usually involve greater or lesser amounts of capital coupled with alternative operating schedules that result in lesser or greater operating costs.

Consider the case where the company actually *does* want to produce 3.124 million m^3/year. This is the five-truck case that was discarded in Table 10.5. Including the two cases already presented, at least five options present themselves:

1. A five-truck match. This option has lower average operating costs than the use of a contractor, but compared to a four-truck fleet the marginal cost of the production from adding a fifth truck to the fleet was uncompetitive with a contractor.
2. A four-truck match, with the additional 300,000 million m^3/year or so of production coming from supplemental work by a contractor. This was recommendation 1 earlier in this chapter.
3. A four-truck match, with the additional 300,000 million m^3/year or so of production coming from overtime work. This will result in higher operating costs, but since there is no more capital (although the existing equipment will wear out more quickly) the reduced "capital" cost may offset the higher operating cost.
4. An underutilized fleet of larger trucks. The alternative considered is a hydraulic excavator coupled with 190-t trucks. The fleet can be worked on a regular three-shifts-per-day roster, with fewer trucks purchased than needed. The trucks that *are* purchased are then well utilized, but the excavator is idle for much of the time waiting for trucks to load. (This is one of two options for having an underutilized fleet of larger trucks.)
5. An underutilized fleet of larger trucks for which the hydraulic excavator is correctly matched with the right number of trucks, thereby achieving optimum hourly production rates. The fleet is used on just two shifts per day or any schedule appropriate to achieve the desired annual production. Because this underutilization is more efficient than option 4, the operating costs will be lower, but capital costs will be higher.

This series of options may seem bewildering. Before computers, such an analysis would not have been attempted. The first two options have already been considered in this chapter. The other three options are all concerned with capital utilization, either underutilization in some way or higher utilization through overtime work.

Figure 10.5 shows the average cost curves for the front-end loader case plotted with an "under-trucked" fleet of larger trucks loaded by a hydraulic excavator.

Presented this way, the case for an "optimized" four-truck fleet of 136-t trucks and front-end loader (supplemented by a contractor) as suggested in recommendation 1 is no longer quite as clear. A three-truck fleet of larger trucks can move about the same annual quantity at about 4% lower overall cost *and* allow scope for expansion. Should the larger equipment be recommended?

In this case the three-truck fleet of larger trucks is superior because it is more efficient and because the annual output from the four-truck fleet of smaller trucks is near maximum for this size of equipment.

Nevertheless, the result raises the question of capital utilization. Are there efficiencies to be gained from better utilization of the smaller equipment? Before a revised recommendation is finalized, the potential for improved capital utilization of the fleet of smaller trucks through overtime work should be

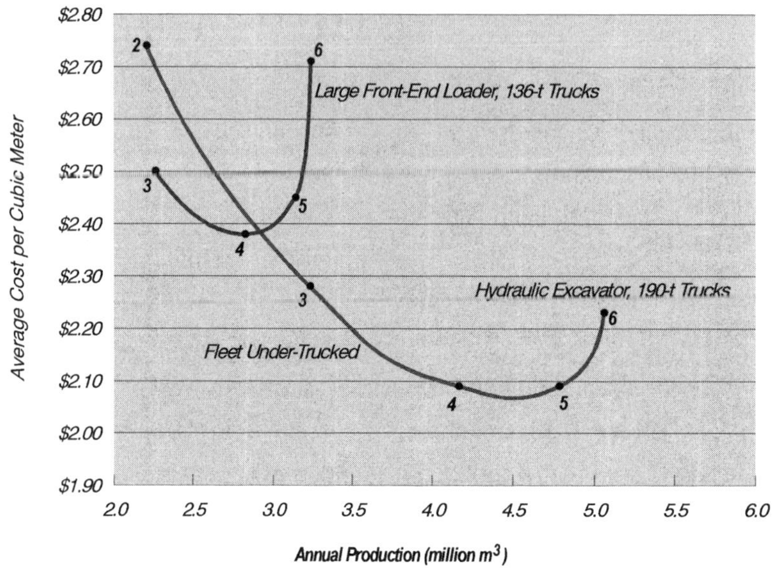

FIGURE 10.5 Average production costs for two fleet sizes

examined. The extended analysis in the next two subsections underscores the value that is potentially lost without thorough research into the alternatives available.

Economics of Overtime

Overtime work increases the *per hour* labor costs but improves capital utilization. Examine again the information in Table 10.1. With overtime work, three changes can be deduced:

1. Operating labor costs for the overtime work will increase. Assuming double time for any work on an extended shift roster, the operating costs for the extra time increase by about 27% for the trucks and about 16% for the loader.
2. Maintenance labor costs will also increase. The maintenance labor costs calculated in Table 10.1 assumed that maintenance would be undertaken during regularly scheduled maintenance times. Once extended shift rosters are implemented, an increasing amount of maintenance will have to be undertaken outside normal scheduled maintenance time. This out-of-hours maintenance is also more costly.
3. Capital costs will decrease. The same capital investment is now producing more output, so capital charges *per unit of production* will decrease. This will be partly offset by earlier wearing out and replacement of equipment, but this too depends on whether the equipment wears out chronologically or based on machine hours.

Table 10.6 shows a recalculation of Table 10.1 for the operating costs of the loader and truck assuming an extended operating schedule with up to 25% additional operating time. In this table, all *operating* labor time beyond the

TABLE 10.6 Equipment operating costs with overtime

Percent Utilization Relative to Base Case	Operating Time (operating hours)	Weighted Average Operating Labor Cost ($/operating hour)	Percentage of Maintenance in Normal Time	Weighted Average Maintenance Labor Cost ($/operating hour)	Total Operating Cost ($/operating hour)	Extra Cost of Extra Operating Time ($/operating hour)
Front-End Loader:						
Base case	4,165	58.43	100.0	30.60	364.78	—
+5	4,373	61.21	95.3	32.04	369.00	453.47
+10	4,582	63.74	90.6	33.48	372.97	456.35
+15	4,790	66.05	85.9	34.92	376.72	459.24
+20	4,998	68.17	81.2	36.36	380.28	462.12
+25	5,206	70.12	76.5	37.81	383.67	465.00
Rear Dump Truck, 136 t:						
Base case	3,329	48.24	100.0	21.44	180.77	—
+5	3,495	50.54	95.3	22.45	184.07	250.14
+10	3,662	52.63	90.6	23.45	187.17	252.16
+15	3,828	54.53	85.9	24.46	190.08	254.17
+20	3,995	56.28	81.2	25.47	192.84	256.18
+25	4,161	57.89	76.5	26.47	195.45	258.20

TABLE 10.7 Marginal costs of overtime

Percent Utilization Relative to Base Case	Operating Time (operating hours)	Average Operating Costs ($/operating hour)	Total Yearly Cost ($)	Marginal Operating Cost ($/operating hour)
Front-End Loader Only:				
Base	4,165	364.78	1,519,309	
Extra	*462*	—	*210,264*	*455.12*
+11.1	4,627	373.80	1,729,573	
One 136-t Rear Dump Truck:				
Base	3,329	180.77	601,783	
Extra	*370*	—	*92,926*	*251.15*
+11.1	3,699	187.81	694,709	
Fleet—Loader Plus Four Trucks:				
+11.1			*581,968*	

annual usage in Table 10.1 is assumed to be at double time. *Maintenance* labor beyond the annual usage in Table 10.1 is assumed to have a decreasing proportion of time within regularly scheduled time. Maintenance labor time outside this regularly scheduled time is also assumed to be at double time. This allocation of maintenance time is also shown in Table 10.6.

From Table 10.6 alone it is possible to deduce the answer to the overtime question. The extra annual production of 312,194 million m^3 means that the fleet will have to work for 11.1% more time. Based on an interpolation from Table 10.6, the marginal operating costs of this additional time are set out in Table 10.7. The extra 312,194 m^3 of annual production incurs an additional

\$581,968 of operating cost, for a marginal unit cost of \$1.86/m^3. There is no additional capital, so there is no capital cost to include. Although there may be a slight increase in discounted hourly equivalent capital cost as a result of earlier equipment replacement, this can probably be ignored for only an 11% increase in hours. Overtime work at \$1.86/m^3 marginal cost competes very favorably with contractor work.

Recommendation 2 (superseding recommendation 1): A front-end loader and four-truck fleet (136-t trucks) may be purchased, with an expected output of approximately 2.8 million m^3/year. If additional production is required, the fleet should work overtime.

Economics of Underutilization

Utilization of equipment is a delicate area of discussion between financial controllers and operations personnel in many mines. Minimization of capital demands maximum utilization of equipment. Yet from an operator's point of view, the availability of equipment to handle peak loads is intuitively worth more than any small gains apparently to be had from squeezing the last hour of use out of a machine.

If experienced operators sense intuitive value, then more than likely they are correct. It remains for the economic analysis to identify where these gains translate into real value (higher NPV or faster payback). This is one objective of this section. A second objective is to alert personnel on both sides of the capital utilization fence that this is an area where intuition can be misleading. Past practices that are the foundation of experience are not necessarily reliable harbingers of the future.

The terms *underutilization* and *overutilizaton* should be used with caution. Custom and practice may have established rules of thumb for deployment of equipment, but from an economic perspective there is no such a thing as *under*- or *over*-utilization. The appropriate utilization is the one that yields the most economical way to move material in the circumstances.

A fully utilized fleet of smaller equipment, including overtime work if necessary, is the current recommendation. The alternative involves larger equipment of a capacity that far exceeds the requirements of the job at hand. The question is: Can this larger equipment be used at lower production rates and still move material competitively with the (technically more appropriate) equipment analyzed in the "Technical Analysis: Waste Removal" section earlier in this chapter?

The initial capital costs and estimated hourly operating costs of this larger equipment are set out in Table 10.8. The two ways to deploy this fleet of larger trucks were described in the "Economics of Capital Utilization" section earlier in this chapter: The fleet can be under-trucked, or it can be "correctly" trucked but used on a reduced shift schedule.

The annual production rates and discounted average cost of production for the first case are calculated in Table 10.9. This discounted average cost calculation,

TABLE 10.8 Capital and operating costs for 190-t truck and hydraulic excavator

Cost Item	Rear Dump Truck (190-t capacity)	Hydraulic Excavator
Initial capital cost, $	1,900,000	5,500,000
Expected life, *years*	8	6
Annual usage, *operating hours*	3,329	4,165
Operating costs per hour		
Operating labor	49.58	55.06
Maintenance labor	24.00	46.40
Fuel	36.18	77.73
Lube, oil, greases	5.43	11.66
Tire wear, replacement	23.52	—
Wear items	9.50	27.50
Repair parts	53.20	99.00
Major overhauls	23.75	55.00
Total operating cost, $/*operating hour*	**225.16**	**372.35**

TABLE 10.9 Average costs of production for hydraulic excavator and various numbers of trucks: under-trucked case

Number of Trucks (190-t capacity) in Fleet	Average Operating Hours per Year per Truck	Excavator Operating Hours per Year	Equivalent Annual Production (m^3)	Discounted Average Cost ($/$m^3$)
2	3,329	4,165	2,204,000	2.74
3	3,329	4,165	3,235,000	2.28
4	3,329	4,165	4,160,000	2.09
5	3,329	4,165	4,786,000	2.10

assuming the excavator and trucks are the only equipment, has been undertaken identically to the example in Chapter 5 (see "A Sample Discounted Average Cost Calculation," p. 66). The costs from Table 10.9 should be compared with the costs for the smaller equipment shown in Table 10.4. A fleet size of three trucks from Table 10.9 is clearly competitive with the smaller equipment.

When the excavator is under-trucked, it spends much of the day waiting for trucks to arrive back from the dump, yet it is still incurring operating cost. Rather than *under*-trucking the excavator, there is another alternative. In this alternative the correct numbers of trucks are assigned for optimum *hourly* production, but the fleet is scheduled for fewer hours per year. This too amounts to underutilization of capital, but at least in this case there are no operating costs incurred when the equipment is not being used.

A series of cases have been analyzed ranging from the 65% utilization level to full utilization. The estimated production and discounted average cost from this series are shown in Table 10.10. Figure 10.6 shows the average costs of production for examples set out in Tables 10.9 and 10.10.

TABLE 10.10 Average costs of production for hydraulic excavator and four-truck fleet: different operating schedules

Percent Utilization	Number of Trucks (190-t capacity) in Fleet	Average Operating Hours per Year per Truck	Excavator Operating Hours per Year	Equivalent Annual Production (m^3)	Discounted Average Cost ($/$m^3$)
65	4	2,187	2,737	2,732,000	2.620
74	4	2,473	3,094	3,083,000	2.420
83	4	2,758	3,451	3,446,000	2.280
91	4	3,044	3,808	3,813,000	2.184
100	4	3,329	4,165	4,160,000	2.090

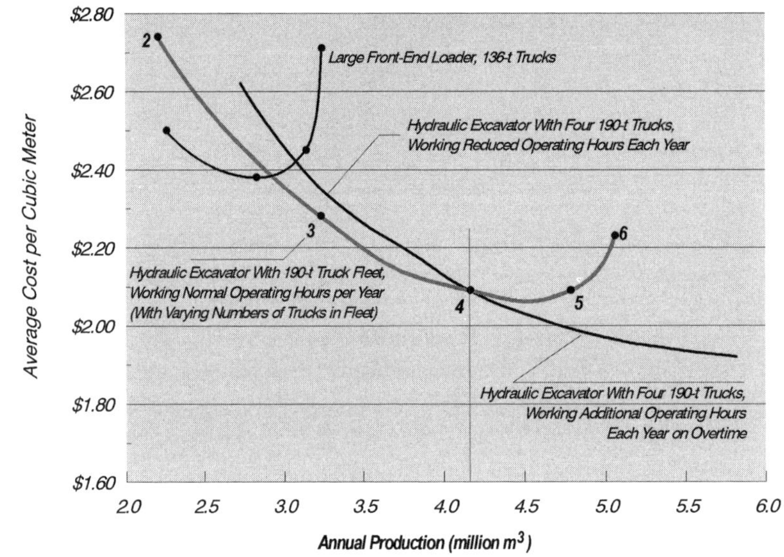

FIGURE 10.6 Fleet average costs: underutilized equipment

At production rates less than 4.16 million m^3/year, the four-truck fleet of larger trucks works a reduced operating schedule. At production rates exceeding this, overtime work is more efficient than over-trucking. As shown in Figure 10.6, reduced shift schedules are not as economical as full shift schedules with a lesser number of trucks operating. Under-trucking is a lower-cost way of achieving the lower production targets.

Is this a general result? Most operators intuitively value operational efficiency over capital utilization, and the result may seem surprising.

The result probably *is* general for trucks and loaders. The reason for this is twofold:

1. In the former case (Table 10.9), there are too few trucks, but each truck is always fully used. Since trucks represent 66% of the operating cost and almost 60% of the capital cost of the fleet, even in this technically inferior case more than 60% of the cost structure is efficient. In contrast, the alternative in Table 10.10 has both the excavator *and* the trucks underutilized.

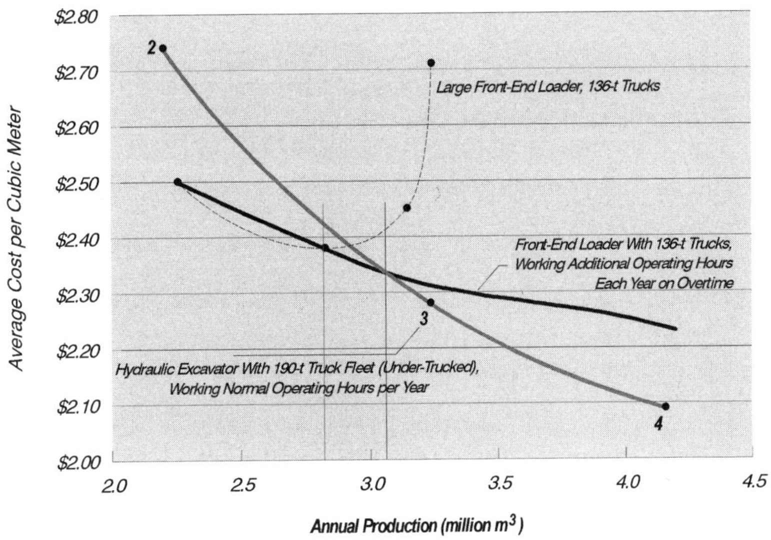

FIGURE 10.7 Overutilized and underutilized alternatives

2. The cases examined are highly capital-intensive. At optimum production, the discounted average cost of production for the hydraulic excavator and fleet of four 190-t trucks ($2.09/m^3) has a capital component of $1.00/m^3. Fully 48% of the "cost" of production is allocated to capital recovery (including return on investment and taxes). Capital-intensive applications are more sensitive to underutilization than are their more operating-intensive counterparts. Most large mining equipment is similar.

Figure 10.7 compares the overutilization of the smaller equipment (by overtime) with underutilization of the larger equipment by under-trucking. Use of overtime allows increased production at lower marginal costs than adding extra trucks to the fleet. However, there is only a narrow margin of production (between 2.8 and 3.1 million m^3/year) where overtime use of a fleet of smaller equipment remains competitive with a fleet of larger equipment.

Recommendation 3 (superseding recommendation 2): If there is no probability that production will exceed 3.1 million m^3/year, then follow recommendation 2; purchase a front-end loader and four-truck fleet of 136-t trucks, and supplement production by overtime work. If production has the potential to expand beyond 3.1 million m^3/year, purchase a hydraulic excavator and only sufficient 190-t trucks to achieve production targets up to a maximum of 4.1 million m^3/year. Beyond 4.1 million m^3/year, operate this fleet of larger equipment on overtime.

ECONOMICS OF OPERATIONAL DECISIONS

Many mines, through imprecise analysis (similar to the example associated with Table 10.5) or though changing mine circumstances, find themselves in a position where they have equipment that exceeds their requirements. This changes the nature of the problem. If there are surplus trucks available, should they be used?

As this section will show, the answer is almost always yes. It may have been a mistake to purchase too much equipment, but it is just compounding this mistake to leave it idle. This result, derived from the economic analysis that follows, accords well with observed practice in mines. Good operators use all of their equipment as much as possible.

Consider again the example in the section "Technical Analysis: Waste Removal" earlier in this chapter where a front-end loader is purchased with five 136-t trucks when in fact the marginal cost of earthmoving by the fifth truck is "uneconomic." When a decision to purchase the fifth truck has still not been fixed, the capital cost of this truck is a variable, and the marginal cost of earthmoving by it has to include a capital component. Once the truck is actually on-site, however, the capital has already been expended, and the marginal cost calculation must exclude part or all of this capital from the calculation of earthmoving cost.

Short of selling the fifth truck (the case covered in the next section), there are two common cases faced by mine operators:

1. If the truck is on-site and likely to stay in the mine fleet for some time, should it be staffed and operated on a regular basis?
2. Most mines have a fleet of trucks adequate to cover their requirements even when an "average" number of trucks is unavailable for maintenance reasons. At various times, there may be *no trucks* on maintenance. Should all of these trucks be operated? What if there is no operator–should one be called back on overtime?

There are many variations of the preceding cases. There is a common solution: Consider *only* the costs and production that will be different whether the equipment is used or not used. Even many *operating* costs are *fixed* costs in such circumstances. A cost is only a variable cost if the person making the decision has the capacity to change this cost within the time frame under consideration. These are the only costs that should be considered.

In the first instance the equipment is on-site, so within the time frame under consideration the "capital" costs will not change. The marginal cost has to consider only direct operating costs. Table 10.11 sets out this calculation, similar to the one shown in Table 10.4. The marginal production is shown in Table 10.5. The marginal costs associated with using the fourth, fifth, and sixth trucks in the fleet, assuming each truck is already on-site, are shown in the table. If the company already has the fifth truck, it is already incurring the

TABLE 10.11 Marginal costs when equipment is already owned

Trucks Used	Annual Production (m^3)	Average Operating Cost ($/$m^3$)	Total Operating Cost Over 1 year ($)	Change in Operating Cost ($)	Marginal Operating Cost ($/$m^3$)
3 trucks in use	2,237,577	1.487	3,327,277		
Marginal cost of deciding to use the fourth truck				595,406	**1.037**
4 trucks in use	2,811,959	1.395	3,922,683		
Marginal cost of deciding to use the fifth truck				604,215	**1.935**
5 trucks in use	3,124,153	1.449	4,526,898		
Marginal cost of deciding to use the sixth truck				605,582	**4.643**
6 trucks in use	3,254,585	1.577	5,132,480		

capital charge, and the (marginal) cost of earthmoving using this truck is still less than the cost of earthmoving using contractors or by scheduled overtime.

A similar calculation applies with personnel availability. Mines typically have personnel in excess of strict requirements to ensure operators are available for major equipment if there is absenteeism. Occasionally, all operators show up and so there are "surplus" operators. Does this make a difference to the calculation?

The logic in this case is identical to the example just given. Consider only the costs and production that will be different whether the equipment is used or not used. "Surplus" operators still have to be paid whether they work or not. The decision to use the truck or not use the truck is therefore just the operating cost *less* the operator's wages for the day.

In this example, the operator's wages represent $48.24 of the total $180.77 of operating cost assigned to the truck. The marginal costs set out in Table 10.11 should be reduced by 27% to determine whether the truck should be used or not. Figure 10.8 shows the results of these calculations, presenting marginal costs as they apply to the time frame under consideration. The marginal costs shown as "Capital + Operating" are the correct marginal costs to apply when capital costs have not yet been incurred and are therefore within the choice set. This is the initial selection criterion that applies to long-term planning because long-term planning personnel *decide* how much equipment to purchase.

For operational planning, capital costs can be regarded as fixed. In the medium-term time frame (6 to 12 months), operating costs including labor costs are under operational management control. Decisions on equipment deployment should be made based on marginal costs for which all operating costs, *including* operating labor costs, are included. (Depreciation is *always* excluded from operating costs for management decision making.)

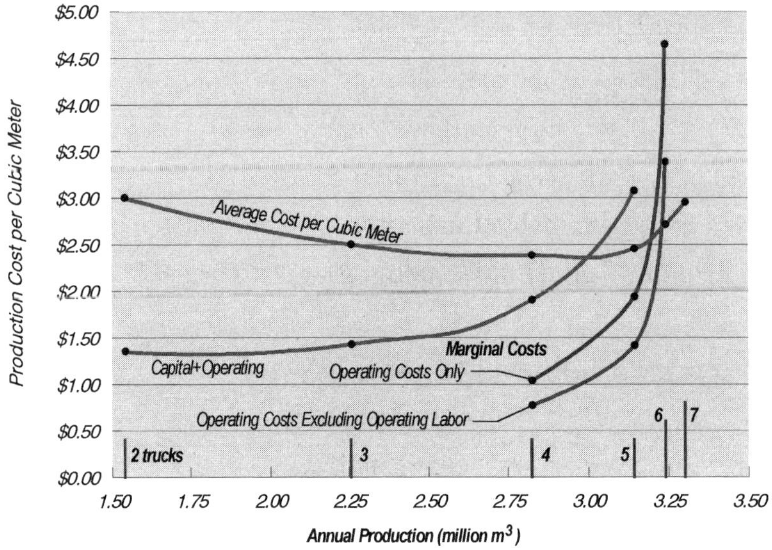

FIGURE 10.8 Marginal costs of over-trucking, with and without operating labor costs

For short-term planning, other operating costs might also be fixed. The costs associated with personnel who cannot be redeployed have to be considered as "fixed" costs, or their costs must be applied only in an amount equal to their opportunity cost in some less important task. With such a high proportion of costs fixed, the marginal costs are low, and there is seldom any case on economic grounds for nonutilization of any productive equipment. If the company has to pay the capital charges and the wages of the operator anyway, then at least the equipment should be used to achieve *some* production.

ASSET MANAGEMENT CONSIDERATIONS

This final section of the case study examines opportunity costs in more detail. It builds upon the preceding section and addresses the following issue: If mine owners have "surplus" equipment on-site, they will probably lose money when they dispose of it. Under these circumstances, would it still be better to keep the equipment?

The easiest answer to this problem is to find out the resale value and conduct an analysis from the start. An alternative approach is to calculate the minimum asking price below which it *is* economical to keep "surplus" equipment.

Consider again the example from the "Technical Analysis: Waste Removal" section earlier in this chapter, where a front-end loader is purchased with five trucks when in fact the marginal cost of earthmoving by the fifth truck is "uneconomic." If the fifth truck has just been purchased, is it absurd for the company to sell it immediately, having now recognized it is not needed?

TABLE 10.12 Capital valuation for equipment disposal

	Production (m^3/year)	Unit Costs ($/$m^3$)	Total Annual Costs ($)
Contractor waste removal	312,194	2.500	780,485
Fifth-truck operating costs	312,194	1.935	604,095
Marginal costs	—	0.565	176,390

TABLE 10.13 Back-calculation of equipment value

	Production (m^3/year)	Equivalent Capital Cost ($/$m^3$)	Total Annual "Capital" Costs ($)
Loader + four-truck fleet	2,811,959	0.981	2,758,532
Loader + five-truck fleet	3,124,153	1.023	3,196,009
Difference	312,194	—	437,476

The value of the truck to the company is determined by the price at which the truck moves material compared to the best alternative price for which the same amount of material can be moved. This "best alternative" price may be one of a number of options. Recall that recommendation 1 suggested that contractors would be a more effective alternative; this was superseded by recommendation 2, which suggested that a four-truck fleet plus overtime was even more economical still. Strictly speaking, recommendation 2 is the option for comparison here; however, for simplicity, assume the minimum price for disposal of the truck is set by the contractor alternative.

There are four steps in the process of this back-calculation:

1. Assuming contractors move the same incremental waste as the (extra) truck, determine the expenditures paid out for this work.
2. Assuming the truck moves the waste, determine the *operating* expenditures paid out for this work.
3. The difference between the expenditures is the cash flow that can be allocated to the "capital" in the truck. Ultimately, this determines the effective value of the truck in this use.
4. Determine how much capital this amount of cash flow will support. Assuming that there are no higher-valued uses of the truck on-site, this is the minimum asking price, below which it is more economical to keep the truck.

The calculations are set out in Table 10.12.

The annual cash flow from Table 10.12 ($176,390) has to be expressed as a capital value, which is not easily achieved since tax considerations also enter into the discounted average cost calculation. The easiest way is to use the equivalent capital cost figures from Table 10.4 and then use this to proportion the value of the truck. The back-calculation is set out in Table 10.13.

In Table 10.13 the difference in capital cost between the two alternatives is simply the annual "cost" associated with servicing the capital in one truck. Since a truck costs \$1.4 million to buy, in this circumstance annual cash flows of \$437,476 are effectively the same as \$1.4 million of initial cash in hand. Any other annual cash flow in the same circumstance can be equated to its corresponding cash-in-hand value based on this relationship. The minimum resale asking price of the truck should therefore be

$$\begin{aligned}\text{minimum price} &= \$1{,}400{,}000 \times (\$176{,}390/437{,}476) \\ &= \$564{,}500\end{aligned}$$

This result—that the incorrectly purchased fifth truck is really worth 40% of its purchase price in this application—*is* quite a general result in these sorts of cases. Detailed analysis of asset valuations built on marginal productivities frequently demonstrates substantial scope for asset management and rationalization. Clearly, assets should be disposed of at the best price available, but the back-calculation set out in the preceding tables provides a powerful pointer to where such rationalizations can yield benefits.

Recommendation 4: If assets are on-site then operators should use them. If the marginal return from using an asset is insufficient to cover the asset's value in the marketplace, then the asset should be disposed of. Assets not disposed of will forever hold back return-on-assets performance indicators. This nonperformance is the fault *not* of operations personnel but rather of the long-term planning personnel responsible for investments and divestments.

CHAPTER 11

Project Evaluation

The derivation of labor and equipment operating costs has been detailed in Chapter 8. These operating costs are usually labeled "direct costs" because they are directly related to the mining operation. Processing and milling costs are also direct costs.

For a complete evaluation of a project or a feasibility study, many other costs must also be considered, including

- administration
- development (exploration and land purchase)
- mine site buildings, roads, and power supply
- freight
- royalties and government charges
- off-site infrastructure, such as townships, road access, power, and water supply

Because these costs typically do not involve large mining equipment and may not be directly controllable by the mine-planning personnel, they have been categorized in this chapter as indirect costs.

This chapter outlines how all the direct and indirect costs that make up a complete mine cash flow analysis are amalgamated.

LABOR NUMBERS AND COSTS

In the developed world, labor costs often represent 50% or more of controllable operating costs in a mine. This section sets out a step-by-step procedure to derive whole-mine labor numbers and costs from first principles. There are three steps in this process:

1. Develop an organization chart showing all site personnel. Production activities that have been separately identified are grouped on the chart by cost center.

2. Tabulate all the personnel on the chart who are not associated with specific equipment or already-defined cost centers. Most technical and administrative staff will be categorized this way. Workforce personnel not explicitly allocated to machines (such as the blasting crew) are also grouped in their appropriate cost center (such as "drilling and blasting").
3. Tabulate all personnel directly associated with each cost center—i.e., personnel associated with equipment already selected.

In open pit mining, almost all members of the operating workforce are directly associated with equipment. Similarly, maintenance labor estimates are strongly related to the mine equipment.

In underground mining, most of the members of the operating workforce are associated with defined activities. Although there may be specific equipment associated with these activities, the equipment operating time may not bear a strong relationship to the number of operating personnel involved in the activity.

For operators directly associated with equipment, the workforce numbers proceed from the equipment operating hour schedule via division by the scheduled time each year that an operator is working the machine. Table 11.1 presents a matrix of required equipment operating time by year, which is divided by the vector of available time for a single machine operator (assumed to be the same each year) to yield the matrix of required machine operator numbers by year in Table 11.2.

The numbers of equipment operators determined in Table 11.2 are not generally integers. In some cases, this can be interpreted as meaning a whole operator will work for part of a year. In some cases, rounding up to the nearest whole number might be necessary during the final tabulation.

The organizational structure of the mine is usually the most logical categorization of labor. The organization chart shows responsibilities and reporting lines and subdivides the organization as a whole into cost centers and divisions that often correspond with cost centers used elsewhere in the mine costing. Typical divisions are

- production
- maintenance
- engineering
- administration
- safety, personnel, and industrial

These can further be subdivided into cost centers. For example, within the production division, cost centers would include

- drill and blast
- waste removal
- loading and hauling of ore (or coal)
- pit service activities (e.g., road maintenance)
- pumping

TABLE 11.1 Labor number calculation from equipment operating hours

Equipment Item	Equipment Operating Time (hours)			Hours per Operator per Year
	Year 1	Year 2	Year 3	
Model C hydraulic excavator	8,300	13,700	13,700	1,750
Model A rear-dump trucks	32,600	48,800	51,900	1,680
Front-end loader	5,000	6,000	6,000	1,750
⋮	⋮	⋮	⋮	⋮
Other equipment	—	—	—	—

TABLE 11.2 Sample equipment operator (numbers) schedule

Machine Operator	Number of Equipment Operators		
	Year 1	Year 2	Year 3
Model C hydraulic excavator operator	4.7	7.8	7.8
Model A rear-dump truck driver	19.4	29.0	30.9
Model B front-end loader operator	2.9	3.4	3.4
⋮	⋮	⋮	⋮
Other equipment operators	—	—	—

The hierarchy of the organization chart integrates the components and allows easy dissection of the costs at each level and/or cost center. In addition, it categorizes personnel, so that it is easy to be sure that all personnel—not just the ones associated with some piece of equipment—are accounted for. A typical organization chart is shown in Figure 11.1. Staff numbers are allocated to each division and section. These numbers vary depending on the size and complexity of the mine. There are no fixed rules. The most reliable aid to judgment is the practice followed in similar operations.

INDIRECT COSTS

Whether a cost is direct or indirect depends on the perspective of the person preparing the cash flow and sometimes on the amount of effort needed for the calculation. There is a tendency to rank the importance of some item according to the amount of effort needed to calculate it. Sometimes very large costs (e.g., freight) are categorized this way. The danger is that more important—i.e., controllable—costs can be overlooked just because they are outside the immediate frame of reference of the person preparing the cash flow.

Assuming that the cash flow is being prepared by a mine-planning person and that anything outside the mining operations area is an "indirect" cost, the following items should be examined for inclusion in the cash flow:

- government charges
- freight
- civil works

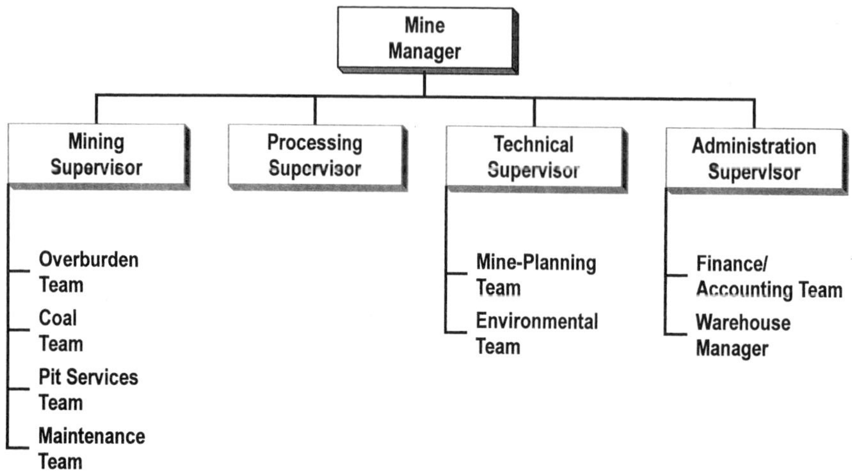

FIGURE 11.1 Typical organization chart

- development costs
- administration costs
- pit and maintenance services

Government Charges

Government charges include royalties and property taxes, as well as levies and bonds for reclamation. They are usually sourced from discussions with government bodies.

Freight

In many mining countries, transport of product from mine site to port is controlled by government-regulated transport methods. In this case, freight charges may not be negotiable and are therefore simple to obtain by contacting the relevant department or contractor. If freight movements involve construction and running of a railway, this railway may call for a very substantial design study and require cash flow analysis in its own right. In such a case, it will be categorized under the civil works cost center or as a completely separate part of the analysis.

Civil Works

Civil works can be subdivided into on-site and off-site works. On-site works include roads, dams, buildings, and power supply; they usually include everything within the mine gate. For a broad-brush study, estimates built up from published civil construction handbooks or magazines are adequate (e.g., a certain dollar amount per square meter or square foot for buildings and another dollar amount per kilometer or mile for roads). For detailed feasibility studies, civil works must be designed and then costed by the appropriate professional group, such as civil, electrical, or mechanical engineers.

Off-site civil works include roads, railways, ports, shiploading, water supply, accommodations, and power supply. For detailed feasibility studies, these features too must be designed and then costed by the appropriate professional group as for on-site costs.

The expenditure on civil works in a large project is commonly very substantial. Detailed assessment of these costs is normally outside the scope of work assigned to mining engineering staff, as well as outside the scope of this text.

Development Costs

Development costs include

- exploration
- land purchase
- the costs of investigations (e.g., groundwater)
- feasibility studies, consulting, and permitting

Cost estimation is usually by quotation. Land purchase costs are usually estimated from valuing adjacent areas. Exploration is usually costed in consultation with the geologist on the project and drilling contractors.

Although these costs are not always a large component of total costs, development costs are high risk—much of this expenditure must be made before commitment when there is no guarantee the project will proceed. Hence, they are closely scrutinized and are often subject to some compromise.

Administration Costs

Administration costs involve head office and site administration before and after project commencement. Items include staff salaries, communications, advice from specialist consultants, plus computers and other office-running costs. These are best sourced from similar operations. Marketing costs are often very substantial in projects supplying a diverse range of products to varying world markets—particularly when a high proportion of offtake is to shorter-term contracts.

Pit and Maintenance Services

Pit and maintenance service items are sometimes categorized as direct costs and sometimes as indirect costs. They include

- light vehicles
- cranes
- general workshop costs
- pumps, lights, and miscellaneous equipment

Capital costs are easily obtained from suppliers, whereas operating costs of small plant are quoted as dollars per week or month rather than on an hourly basis.

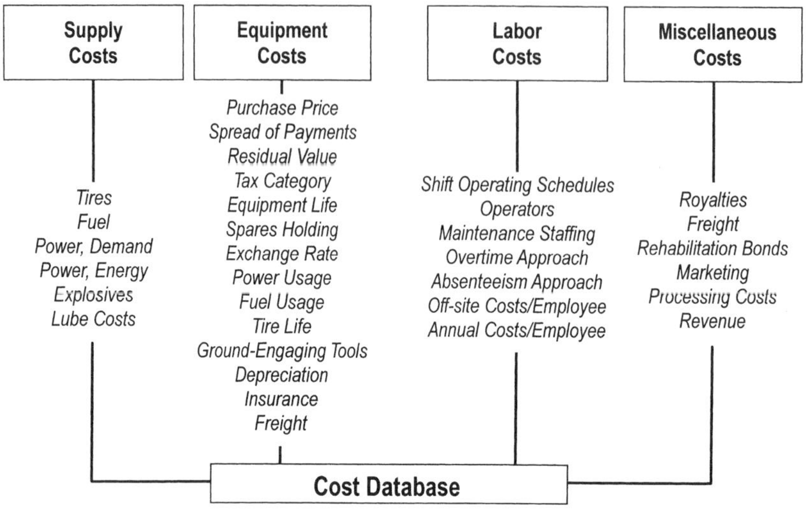

FIGURE 11.2 Cost database

BUILDING UP A CASH FLOW

This section addresses the issues associated with building up a cash flow, including a cost database, cost centers, and the process of taking data from the original mine production schedules up to tabulating the final cash flow.

Cost Database

Cash flows are usually derived by multiplying operating items (such as fuel and explosive quantities used) by their unit costs. The table of unit costs constitutes the cost database. Figure 11.2 shows the typical input and the structure of this database in diagrammatic form, which includes

- supply costs
- equipment costs
- labor
- miscellaneous

The derivation and source of these costs have already been discussed.

Cost Centers

To rationalize cost estimation, whole-project studies should be divided into cost centers, or groups of activities related by function. Within each cost center, a budget estimate can be determined to include

- equipment operating costs (further divided into fuel, consumables, and materials)
- labor
- capital expenditure

TABLE 11.3 Typical whole-mine cost centers

Cost Center	Component	Cost Center	Component
Development	Exploration Land purchases Feasibility studies	Administration	Head office Management Engineering Stores
Waste removal	Ground preparation Drill and blast Overburden Parting removal	Ore mining	Drill and blast Grade control Ore mining
Processing	Crushing Stockpiling	Pit services	Dewatering Road maintenance Reclamation
Maintenance services	Cranes Fuel and service facilities Workshop costs	Surface facilities	Site buildings Site power and water supply Mine roads
Freight, royalties	Rail and port handling Sampling Government royalties	Off-site infrastructure	Township Roads Power and water

Typical cost centers are set out in Table 11.3.

After calculation of costs, the presentation of costs by cost center can be made with pie charts. Figure 11.3 shows an example pie chart from one mining study.

The whole-project costs can be presented in more meaningful form showing only the costs pertaining to a particular set of cost centers. Figure 11.4 shows the components of the costs from Figure 11.3 pertaining to four mining cost centers (see the pie chart on the left half of the figure). Within this *mining-only* cost grouping, the dragline cost center is shown (the bar on the right side), comprising the equipment and activities associated only with this cost center.

Flowchart

Keeping track of the thousands of elements in a complete planning study is complex, and there is a danger of completely missing some important cost. Figure 11.5 shows a flowchart for systematically building up all the elements of the final cash flow from the component schedules.

The steps in the preparation of this final cash flow are as follows:

1. The production schedule is prepared, with all major plant components itemized, the shift configuration specified, and all production quantities and quality characteristics (by time) determined.
2. The schedules of major operating items, such as equipment operating hours and usage of operating supplies (consumable quantities), are prepared.
3. The project is divided into cost centers.

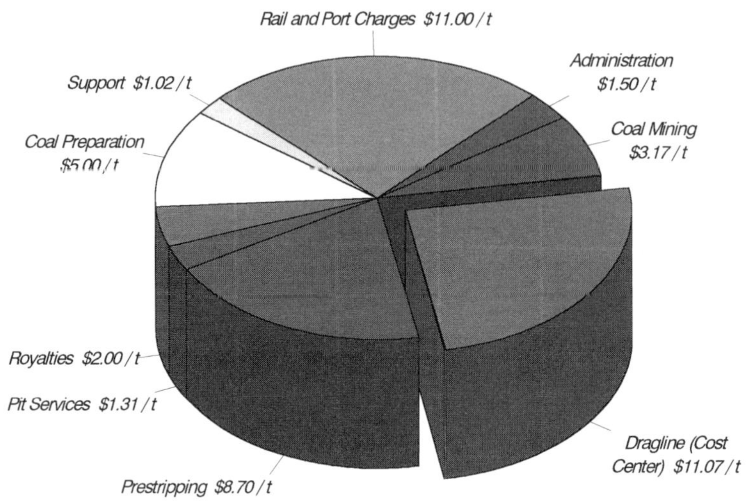

FIGURE 11.3 Whole project costs by cost center

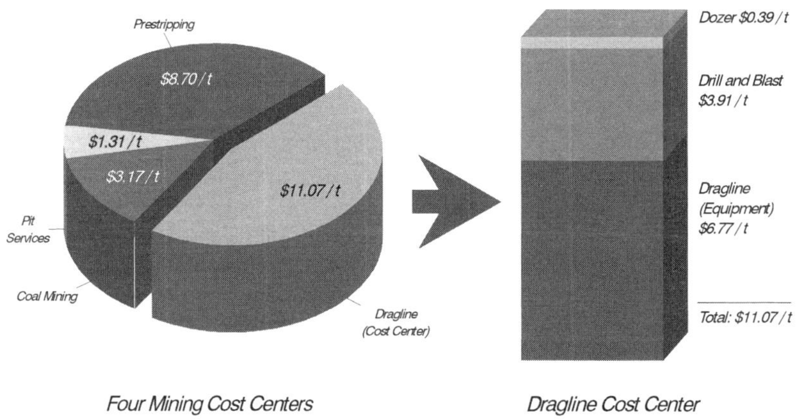

FIGURE 11.4 Subdivision of cost centers

4. Equipment purchase and replacement schedules are determined either directly or from the operating hour schedule. Labor numbers are calculated similarly.

These form the major input to the costing schedules. These operating schedules are multiplied by the unit costs from the cost database to calculate

- equipment operating costs
- labor costs
- total operating costs
- total capital costs
- cash outflow

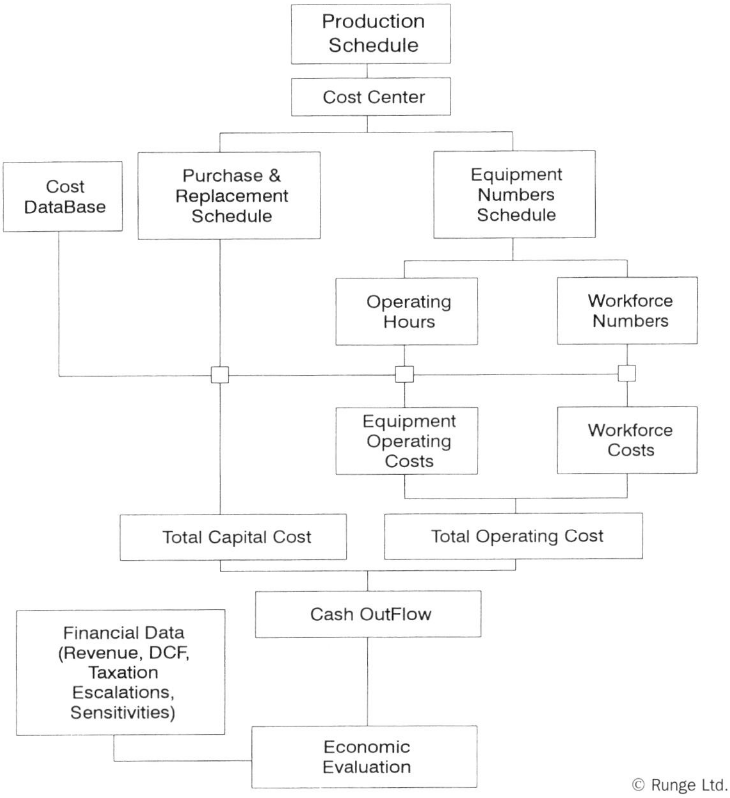

FIGURE 11.5 Flowchart of mine-costing procedure

The procedure for this tabulation is analogous to the procedure in Tables 11.1 and 11.2.

After these schedules of annual costs are prepared, a more detailed economic evaluation (or financial evaluation) can be undertaken where additional inputs are incorporated, including

- revenue estimates and future market conditions
- taxation, alternative schemes that may save or delay taxes, and depreciation
- escalation, as well as likely changes to the plan if characteristics of the mine change
- interest rates, debt/equity considerations, and corporate structure
- strategic risk assessment
- sensitivity analysis

This cycle may have to be repeated many times before a project reaches the implementation stage.

The final set of discounted cash flows is treated as described in the examples set out in Chapters 9 and 10.

CHAPTER 12

Mining Strategy

The previous 11 chapters have detailed how mining choices are evaluated in economic terms by using conventional analysis tools. These conventional tools usually restrict analysis to the project immediately at hand, and they assume that the alternative to project approval is a risk-free market-determined opportunity, such as treasury bills or long-term government bonds. More important, conventional analysis assumes a passive business environment or, at the most, an environment that is changing in predictable ways.

The strategic approach to mining decisions assumes an active environment. Decision makers in this environment are surrounded by other active decision makers. Different choices are interdependent and recognized as such. Choice is influenced by the decision maker's *expectations* of the actions and reactions of others.

This chapter introduces mining strategy. It highlights the risk and return dichotomy, as well as how investment evaluation tools designed for low-risk business decisions (most of the tools and techniques discussed in the previous 11 chapters) can sometimes lead to erroneous choices. In addition, it questions the way mining practitioners actually make decisions, and it suggests alternative ways of applying the decision rules for strategic choice.

INTRODUCTION

Strategic thinking is best illustrated through an example. (This example has been paraphrased from a similar example in Dixit and Nalebuff [1991].) Think of the difference between the blasting supervisor in a mine and a general on the battlefield. When setting out a plan, the blasting supervisor hardly expects to come back the next day to find the ore has moved (hidden itself) because it did not want to be blown up. The planning environment is neutral. The process is a mechanical one, and the outcome is independent of the actions or reactions of players unconnected to the choice. A general setting out plans in a war zone knows the enemy will be trying to anticipate his actions. The general's plan must incorporate their expected reaction *and* how that reaction, if it materializes, will be accounted for. Decision making in the business world is similar to decision making on the battlefield. Customers, competitors, and suppliers are intelligent and purposeful people whose aims cannot be assumed consistent with the aims of a mining company seeking to maximize wealth for its shareholders. Many of the assumptions in a mine plan are dependent on the actions and reactions of these other participants, and the impact of their likely changes must be taken into account.

For many decisions in operating mines, the standard (nonstrategic) approach is quite appropriate. If the choice is between an electric rope shovel and a diesel-powered hydraulic excavator, this choice–if made on the assumption of an unchanging world–is unlikely to be wrong even when the outside world changes quite substantially.

However, for many decisions the assumption of changelessness is inappropriate. Even if these influences are outside the control of mine operators, the likelihood of change must still be taken into consideration–particularly for whole-project analysis.

The balance of this book examines decision making within such a dynamic environment. Evaluation tools designed for unchanging environments frequently lead to loss of project value when applied incautiously to dynamic environments. The new evaluation tools and new ways of using existing evaluation tools aim to redress this problem.

The primary question (and focus for the balance of this text) is the one directly posed by P.F. Drucker (see Drucker [1970], quoted on p. 59 of Burgelman and Maidique [1988]; emphasis added): "Decisions exist only in the present. The question that faces the long-range planner is not what we should do tomorrow, it is: What do we have to do *today* to be ready for an uncertain tomorrow? The question is not what will happen in the future. It is what futuristics do we have to factor into our present thinking and doing; what time spans do we have to consider, and how do we converge them to a simultaneous decision in the present?" Once a project has been proposed or is in operation, how can it be structured or managed for the highest probability of achieving expectations?

One caveat–the techniques in the balance of this book are not as analytically well developed as the techniques in the preceding chapters. There are two reasons for this:

1. The world of human interaction is not as clear-cut as the world of machines. It is certainly not as amenable to robust analysis. The first 11 chapters of this book are primarily working in the world of machines, whereas the strategic approach draws much more heavily on models of human behavior that are not as predictable. Results are less robust.
2. With minor exceptions, the theories of finance that underpin the decision models in the first 11 chapters are well developed. This is not so for choice in a dynamic environment. In this case, economic theories of decision making under uncertainty, as well as the branch of economics that deals with strategy (i.e., game theory), are much less developed.

Thus, some of the tools described in the following pages are useful only in a conceptual way, not as analytical devices for quantitative results. Further, capital investment choice under uncertainty (the model set out in the balance of this chapter) is still subject to testing in economics and psychology, and with the inexorable advance of knowledge this too may change. Few empirical studies have been completed on the relationship between investment and uncertainty, and what work exists is far from conclusive (Leahy and Whited 1995).

In one sense, a capital investment choice is no different than any other choice. The value of one alternative is compared in the mind of the decision maker with the value of the next most evident or attractive alternative. In another sense, there is a world of difference. In the consumer world, value is more evident and usually realizable immediately. The capital investment decision process, on the other hand, is distinguished by two characteristics:

- The "value" in capital investments materializes only in the *unknowable* future. It takes time to come to fruition. The choices in a capital investment framework are choices among *strategies* leading to the future, and these strategies themselves involve follow-on choices not wholly under the control over the decision maker (Arrow [1958]; see p. 61 of Arrow [1984]).
- Capital investments are heterogeneous bundles of value. Consumer goods can be tried and returned with little loss of value, and choice is informed through the benefit of repeat purchases. The second-hand value of a mine shaft built to access an unknown orebody cannot readily be returned for money back or turned to some other purpose.

It is not surprising that the economics literature, largely built on the assumption of homogeneous commodities in perfect markets valued with costless perfect information, falls short of satisfactory explanations for the capital decision process. This chapter sets out an alternative model.

One terminological point is in order. In the balance of this text the term *project* is used in a very broad sense. It represents one of the strategies leading to the future—a strategy involving capital investment or divestment. A project such as the planned purchase of a truck and loader fleet defined in the physical sense may lead to several logical "projects" in the context of this and the following chapters. In this logical definition of a project, project A might mean "buy a new truck fleet tomorrow," and project B might mean "buy a new truck fleet at the start of the next financial year." The examples illustrate very different treatment between physical and logical projects, particularly when risk is explicitly considered.

The tools throughout this book define the mechanisms to compare projects, but there is a precursor step to these analyses. Someone has to conceptualize the projects to be analyzed. Once a project has been defined, then comparison with alternatives can proceed, but where is the greater scope for incorrect choice? Is it through incorrect or inappropriate mechanisms for comparison, or is it through failure to conceptualize choices that are fundamentally of higher value? The greatest contribution that can perhaps be made by a practitioner aiming to maximize economic value from mining-related choices may be the value in the extended possibilities that otherwise escape consideration, rather than the better analytical treatment of a smaller set of alternatives.

This notice bears repeating. Within an organization, decisions are usually arrived at through a well-defined procedure that eliminates many alternatives from consideration, and frequently some of the eliminated alternatives represent higher value than choices finally adopted. The issue of how an organization "knows" whether it knows enough has been addressed by a number of authors. Runge (2000) examines path dependency in organizational decision making and demonstrates how the shifting risk or return focus introduces a bias in the process. Shackle (1983) examines knowledge, or lack of knowledge (what he called *unknowledge*), from a broader perspective.

What is clear, through observation, and with the benefit of hindsight, is that many poorly selected choices are made by mining companies who did not know that they did not know some important fact, yet the knowledge may have been available within the mining industry or even from within their own organization at the time the decision was made. This conceptualization—an entrepreneurial understanding of extended possibilities—though vitally important, is excluded from the scope of analysis. Additional comments on the alertness to extended possibilities and promotion of entrepreneurship are included in Chapter 16.

THE INVESTMENT DICHOTOMY: RISK AND RETURN

Mining has always fit uneasily into any standard industry investment models. Each mine deals with unique and wasting assets. Uncertainty is substantial and frequently unresolvable. Rules that apply across a broad industry spectrum are commonly overlooked or subject to special exemptions in mining.

A further characteristic of mining is notable. Industrial enterprises sell products with well-defined characteristics into markets for which a market value can be readily assigned—at least, this is the assumption in typical economic models. The markets for mining products, on the other hand, are far from well defined. Geological characteristics are different in every deposit. Ores and intermediate-stage products contain different amounts of impurities that themselves often have value to particular customers. Government controls and strategic stockpiles have strongly influenced price, at least until the fairly recent past. The fully competitive equilibrium that is assumed in much economics work does not characterize many mineral commodity markets.

Since the mid-1980s, increasing competition and internationalization have led to a substantial move toward this market ideal. Rigidities have been reduced. Government interference in the marketplace has diminished. In major mining companies, this trend has been accompanied by treatment of mining investments no different than that which companies in any other industry apply to their investments. Return on investment and shareholder value are paramount, and any unique characteristics are assumed to be minor.

This trend bears reexamination. Risks are not unique to the mineral industry, but it is in the area of risk that mineral industry investments differ most from standard investment models.

In stock market investments, diversification reduces risk and therefore makes sense for investors. Within a firm, the same is true. A firm with only one mine centered on one commodity will have returns that are more volatile than a similar sized firm with three (smaller) mines producing three different commodities. Yet this style of firm may not be as attractive to investors as a nondiversified one. The reason is that investors themselves can diversify–putting one third of their assets into each of three different firms who focus on one commodity alone. Economic theory usually regards this approach as much more efficient because investors can move their funds from company to company at less cost than a company can buy and sell mines and processing plants.

The validity of this theory extends only so far. The limitations relate to the uniqueness of orebodies, the nonreplicability of the mining processes, and the extent to which private knowledge adds value that generates synergies.

1. Where production processes have no unique inputs, a firm producing at one-third the capacity will have higher costs of production than a firm producing at three times the rate. In the mining industry, each mine has a unique input–the orebody–that influences the cost of production. A small orebody extracted at low production rates can compete with other, much larger producers.
2. Different ores occur together. Silver is produced in conjunction with lead and zinc mining. Copper occurs with gold and molybdenum. Companies mining these ores have automatic diversification mechanisms at lower costs than can be provided through market mechanisms.
3. In the stock market, knowledge transfer among participants is virtually instantaneous and relatively costless. Thus, the learning effort when resources are reallocated between investments can be largely ignored. Conversely, within a mining company as orebodies are worked out and new orebodies opened up, knowledge effects are quite significant. Risks as faced by mine operators are idiosyncratic and cannot be diversified away.

Thus, the application of conventional models to understand risk and return, and the actions that follow from them, must be made cautiously. With financial investments, opportunities exhibiting low risk and low return form part of a continuous spectrum through to higher risk and higher return, and the choice of a higher-return opportunity is synonymous with the choice for higher risk. In mining, the uniqueness of orebodies and the idiosyncratic risk attached to exploiting them means that projects offering higher return do not necessarily involve higher risk. This creates opportunities that make the industry much more interesting. It also inhibits some decision making when projects are evaluated by using "efficient market" models but have risk/return characteristics inconsistent with these models.

As the transition from a less competitive era to the more competitive era evolves, the rules of evaluation also require adaptation. The criteria and model refinement in the balance of this book are aimed at accelerating this adaptation process. With rules in transition, high-risk decisions will continue to be made (unknowingly) in the expectation of greater returns, and "acceptable risk" projects will be passed over because the risk is not understood.

CRITERIA FOR DECISION MAKING

This section sets out the conceptual tools to rationally compare choices subject to differing risk and return characteristics.

If a past decision has yielded results different than original expectations, then this could be due to any combination of three causes:

1. Luck (good or bad).
2. Insufficient or incorrect data.
3. A flawed process of *recognizing* value.

(There are additional causes, most notably if one or more participants in the decision process are following an agenda that is inconsistent with corporate objectives. These "agency" issues are overlooked in this text.) One difficulty in any analysis is that successes and failures are judged with the benefit of hindsight. From this perspective the information upon which the decision was made is lost among the new information made available as the project unfolds. A manager who through pure luck started a nickel mine just when the nickel price peaked (and then sold forward at maximum price when the market fell) would be a hero. One whose luck went the other way would be a pariah. Although luck is surely not an unimportant element, it is of limited benefit in this discussion because even if the exact circumstances were revealed and understood after the event, that knowledge would not help for future decision making.

Luck cannot totally be excluded, though, since even if some unknown characteristic is varying in a totally random way, effort and discernment are still necessary to establish this fact. Stigler (1961) has examined the economics of information, and in cases where unknown information can be assumed to follow some established statistical model, has demonstrated an optimum amount of search. Strydom (1986) has examined information in a totally different context by using a subjectivist approach consistent with the genuine uncertainties similar to many mining problems. The following example illustrates this problem and introduces the first issues leading to a rational process of decision making.

Example 12.1:

Suppose you are presented with a business opportunity akin to the casino game of roulette. Suppose also that there are lots of things about this business opportunity that you know you do not understand. What is a rational way to size up this opportunity before investment?

Evaluation processes start with some research. Most actors will pause for a time and observe the game. In this example, after 30 spins of the wheel, the ball drops 26 times on red and 4 times on black.

If this was the basis of your knowledge and you analyzed it by using the efficient market model or by case studies, when you entered the market you would choose red and expect 26 wins out of the next 30 throws. Assume now you won 15 out of the next 30 throws. Your investment would have been unsuccessful.

Clearly when we make decisions we are not so naive as to blindly extrapolate past trends. We always apply *some* knowledge of the process. But how much knowledge do we need? A "little" knowledge of statistics would tell you that "on average" there is a 50/50 chance of red *or* black. Therefore, having just witnessed a run of 26 red out of 30, you could easily assume that for the next 30 spins there is every expectation of 26 *black* and 4 red. You would choose black. You would also look askance at your naive competitors who are betting on red. And again when you won only 15 out of the next 30 throws, you would have to declare your investment as unsuccessful.

The problem is how to separate the uncertainties that are random from those that are conceptually predictable, and this can be done only by understanding the process. Gambling is a zero-sum game, so anyone that understands this process would not play. On the other hand, business generally *creates* value, so even if we enter into transactions like this example we might still come out averaging more than 50% wins no matter what side we took.

The future might be largely unknown but it is not *unimaginable.* With current decision-making processes, much about the future that *is* known does not get a chance to be used. As Wack (1985) observed in an illustrative example: "Suppose heavy monsoon rains hit the upper part of the Ganges River basin. With little doubt you know that something extraordinary will happen within two days at Rishikesh at the foothills of the Himalayas; in Allahabad, three or four days later; and at Benares, two days after that." This is a prediction, not a forecast. And, as Senge (1990a, p. 320) highlighted: "[A prediction is] something you can say with confidence about the future, because it depends not on projecting historical data into the future, but on understanding the dynamics of an underlying system." Many observers would not draw such a fine line between the words *prediction* and *forecast,* but the point is clear. An extrapolation of a past trend is categorically different than an assessment of likely future conditions made with an understanding of the dynamics of the underlying system. Predicted future conditions may contain uncertainty, but the presence of this uncertainty should not obscure the real value of the expected outcomes for longer-term planning.

The balance of this section looks at how rational decision makers go about understanding the value in choices subject to uncertainty, leading to the adoption of one alternative. An example is again an instructive starting point.

Example 12.2:

Envisage first a situation at a mine site where additional equipment is needed. Perhaps the inventory of drilled ground is declining fast and a new drill is required as soon as possible. Current inventories mean that production will not suffer for at least 6 months, but this is the lead time for a new drill to be purchased, delivered, and operating. To a planning engineer, the need is

> urgent. Others in the organization—the drill supervisor, the mine manager, the company board—have different concepts of urgency and timing. Small changes in the production from existing drills can delay the critical point for perhaps 18 months. "Uncertainty" is a very ill-defined characteristic among the participants in the decision. Yet the decision requires the motivation of all of these parties.

The preconditions for decision making (necessary prerequisites of human action) as paraphrased from Mises (1966) and Locke (1995 [1693]) are as follows (all three are necessary):

1. Dissatisfaction or uneasiness with the current state of affairs or with the state of affairs as they are envisaged to materialize in the absence of action.*
2. An imagined set of conditions that are more satisfactory.
3. The expectation that purposeful behavior directed toward bringing about the imagined set of conditions has the power to remove or at least alleviate the felt uneasiness (Mises 1966, p. 14).

These three conditions could be put into dry economic terms. The first condition spells out the net present value of the "do nothing" alternative. The second condition spells out the net present value of the proposed course of action—presumably being greater than the "do nothing" option. The third condition, an *expectation,* emphasizes the *uncertainty* aspects of fitting the proposed course of action into an existing framework. (There should also be uncertainty attaching to the "do nothing" alternative. In many cases, the uncertainty associated with proposed new courses of action is defined by default as the *additional* uncertainty beyond the status quo.)

Action is therefore the outcome of choice among alternatives, each of which has an *expected value* and each of which has some *uncertainty* that this value will materialize. Under uncertainty, "value" has at least two dimensions. In a corporate context, a choice by all parties to the decision relies on a meeting of the minds in four areas: (1) the expected value of the "do nothing" option; (2) the uncertainty surrounding this option; (3) the expected value of the proposed alternative course of action; and (4) the uncertainty surrounding this alternative course of action.

A further characteristic frequently overlooked or simply assumed is the mechanism for formulation of alternatives that facilitate choice. On the one hand, *default alternatives* are self-evident. Nevertheless, to be rationally evaluated, they still require some expectational mechanism. The participants have to envisage in their minds the future set of conditions that will likely transpire in the absence of action. On the other hand, *new alternatives* involve a twofold and more difficult expectational problem. First, participants have to envisage something that does not exist but that through their actions might conceivably

* Locke says: "What is it that determines the will in regard to our actions?...[It] is not, as is generally supposed, the greater good in view, but some...uneasiness a man is at present under. This is what successively determines the will, and sets us upon those actions we perform" (Locke 1995 [1693], p. 176).

be brought into existence. Second, they have to envisage how this alternative might fit into the existing framework and the future set of conditions that will likely transpire in this framework. This is a nontrivial and frequently overlooked issue.

"VALUE" WITH UNCERTAIN CHOICE

Decision making and implementation are always directed toward the future; they are essentially and necessarily always planning and acting for a better future. Their aim is always to render future conditions more satisfactory than they would be without the interference of action. The uneasiness that impels a person to act is caused by a dissatisfaction with expected future conditions as they would probably develop if nothing were done to alter them.

For *operating* mines, changes have to be made as the mine and market conditions change. The success of any *new* investment is primarily a function of how successfully these ongoing changes are made. In an assessment of what value to assign to the original investment—before commencement—one key consideration is how easy or difficult it will be in practice for these ongoing changes to occur. A key to capturing this value is the degree of sophistication of reporting and alertness of operators as to what is important or not important. These perceptions cannot be assumed.

Example 12.3:

Declining pit inventories or floor stocks usually signal problems to a mine operator, yet accounting personnel often regard the declining stocks as a welcome reduction in working capital. Smaller floor stocks mean less flexibility in the mining operation and higher costs due to less efficient equipment deployment. Without an objective mechanism to quantify (in accounting terms) the effect on the mine *as it would probably develop if nothing were done to alter it,* then this consistency will be missing and necessary change will be delayed.

In a world where mines change slowly, trial and error can act as the mechanism to align value among disparate participants. In a world of rapid change, trial-and-error methods are too slow—by the time the error has been recognized, the circumstances surrounding the next trial are vastly different than those of the last. The complexity of modern mines means that operators have only limited intuitive ability to foresee very far into the future.

In assessments of alternatives, effort can be expended in understanding any of the four areas (see the preceding section of this chapter) that underpin choice. The payoff from effort is not equal for each of these four areas. If the "do nothing" alternative and the preferred new course of action are each quite uncertain, then this uncertainty has to be addressed first. Companies have different capital structures and different tolerances for risk for any one type of investment, and the first priority is to ensure that projects fit within this tolerance.

Once the uncertainty is resolved to within these corporate bounds, efforts to further reduce the uncertainty yield poor returns. Growth (i.e., real economic value added) is an outcome of maximizing the difference between the expected return of the proposed new course of action and the default case, not of a reduction in uncertainty. This should be the primary focus of effort for the long-term health of the enterprise.

How, then, do rational actors formulate value and make choices in an envi ronment of uncertainty—allocating resources between uncertainty reduction and NPV enhancement? (A comprehensive theory of large capital investment choice has been developed by Runge [2000]. This section is an abridged form of this theoreticial development.)

The objects of choice in a "capital" framework are alternative *strategies*. Actors are choosing between one path into the future (and the complete future opportunity set implied with that path) and some other path. Each strategy includes desirable and undesirable attributes subject to uncertainty, the fulfillment of which is at least partly influenced by the actors themselves. Choice between alternative strategies implies a mechanism to comparatively rank these attributes, including the scope to influence their fulfillment.

Actors in this environment do not simultaneously "weigh up" and trade off risk *and* return. Actors first satisfy a risk criterion, or uncertainty-based criterion (in an expectational sense); this is a precondition to any assessment of value based on the NPV criterion. "Risk" in this environment is relative risk. An actor who chooses to go mountain climbing instead of staying home and reading a book (the most attractive alternative foregone) is taking a risk only by way of the *difference* in potentially unfavorable outcomes between the two alternatives. Thus, *real* risk requires choice.

Choice alternatives that satisfy this uncertainty precondition are then valued according to the expected NPV.

Figure 12.1 shows the characterization of risk and uncertainty. Uncertainty is represented by the distribution of values. Value is represented on the horizontal axis and the relative likelihood of occurrence on the vertical axis. The reference line (NPV = 0) represents the point of indifference and corresponds to the return when discounted at the cost of capital. (If the distribution were plotted with "return on investment" on the horizontal axis, the reference line would be drawn at the marginal cost of capital.) The outcomes for which the value falls to the left of the reference line represent risk.

This model of large capital investment choice contrasts with most models of decision making elsewhere in the economics literature. The theoretical development of the model has been set out elsewhere (see Runge 2000), but it is instructive to examine the implications of the model and the support for it in a mining industry context.

For example, mine planners frequently examine their different plans to a high degree of detail, refining their costs to perhaps ±5%, only to find that in a short time the selling price of their product has changed by many times this amount. Does the refinement of costs to this level of detail make sense? The refinement of costs *does* make sense under this model if it is a point of differentiation

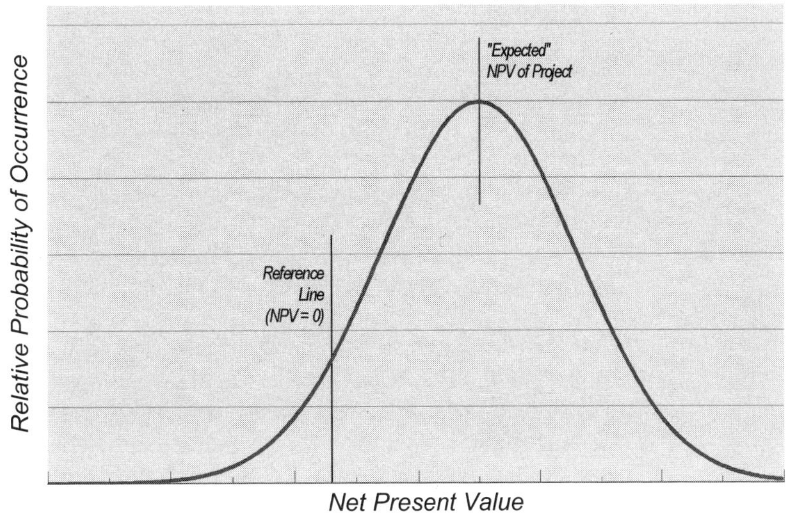

FIGURE 12.1 Distribution of values, risk, and uncertainty

between alternative ways to develop the mine. It is also *quite* rational under this model to give the "selling price" scant consideration *provided that* changes in this price affect all alternatives equally.

Procrastination—very commonly observed in this type of choice situation—is also rational under this model. Procrastination in common parlance means "delaying making a decision." It is supposed to be a failing of management. In this model, the decision to delay is also a choice. An alternative way of describing procrastination is as "choosing to proceed but with an option to abandon if certain conditions do not happen in the immediate future." The "certain conditions" relate to resolution of uncertainty. The "do nothing" alternative might have a low NPV, but it might satisfy the uncertainty criterion (within a company's tolerance for risk), whereas the alternative with a notionally high NPV may not. If uncertainty is likely to be less in the future, then the present value of this alternative, when assessed *then,* may still exceed the present value of the status quo.

Some of the great mines of the world have come about only through a tortuous sequential process consistent with this model.

Example 12.4:

Suppose there is a style of mineralization that is very variable. If you knew exactly what was in the ground and how to mine it, the after-tax real return on investment would be 20%. (Assume for this example that an internal rate of return of 20% is very high and an internal rate of return of 5% is very low.) Unfortunately, you don't know the orebody characteristics, and if you properly researched them to satisfy your company's tolerance for risk (say, less than 10% chance of a return less than the cost of capital) then the cost of this research would reduce the return to only 5%, making the project nonviable. You could drill out just a portion of the orebody and start at a small production rate, but use of this smaller equipment also reduces the return to only 5%. How can such an orebody be developed?

The classic answer to the type of problem posed in the preceding example is diversification. A portfolio of like orebodies—not geologically similar but similar in terms of difficulty and cost in resolving uncertainty—can be developed simultaneously. The return from the unexpectedly bad projects in the portfolio will balance the return from the unexpectedly good projects. On average the expected return will be achieved, and the risk *in aggregate* will be within the company's tolerance for risk.

What if a company is confronted with just one very large project of this type? If history is a guide, such projects are developed by a series of companies attempting development but often going broke. Each attempt at development leaves subsequent owners with a legacy of knowledge that eventually is sufficient to allow large-scale profitable exploitation within the risk tolerance of an owner. In the oil industry, the same problem occurs with individual high-risk deep wells for which uncertainty is irresolvable in advance. Large oil companies solve the problem again through diversity. An investment with 10 oil companies each owning one-tenth of a portfolio of high-risk wells yields the same expected return as a single well developed by each company alone. The aggregate investment satisfies the uncertainty criterion that would otherwise inhibit the sinking of *any* wells.

The balance of this book applies this model of choice to examine a range of mining problems. The model is applied in stepwise fashion.

1. The first step concerns the potential for outcomes for which the return is less than the cost of capital. This benchmark acts as the precondition to direct subsequent action or choice.
2. Among choices that satisfy the precondition, energies are focused on enhancement of expected NPV, and alternatives are chosen on this basis.
3. If choices do not satisfy the precondition, additional information is sought until the precondition is met. If some but not all choices satisfy the precondition (but the ones *not* satisfying it have higher expected NPV), then delay is rational.
4. When the uncertainty precondition cannot be satisfied among *any* choice alternatives, "value" is determined by the case that has the least risk, regardless of the expected NPV.

Chapters 14 and 15 explicitly examine choices using these conditions.

CHAPTER 13

Traditional Mechanisms for Valuing Mining Enterprises

This chapter discusses the traditional techniques used for evaluating mining projects, including short-form techniques used by operating mines and techniques for formulating mining strategy in a corporate-wide sense.

TRADITIONAL PROJECT EVALUATION

New projects are normally planned in a head office environment by project teams specifically dedicated to that purpose. Operating and capital costs are developed reliably from quotes and operating experience. Several mine plan scenarios are developed to examine the sensitivity of the project to different planned production rates.

The tool used for these evaluations is almost exclusively the discounted cash flow technique. Most companies also have rules of thumb either derived independently or adapted from discounted cash flow analysis to compare the project with other projects in their experience. Sensitivity analyses are undertaken to determine the sensitivity of the return on investment (or net present value) to changes in the input parameters. Sensitivity analyses are undertaken on product selling prices, some production rate variations, and other key inputs affecting costs or revenues (e.g., exchange rate, capital cost changes, fuel oil prices, delays in revenues after start-up).

This mine evaluation process—feasibility study, discounted cash flow analysis, sensitivity analysis—leads to a high degree of confidence in decision making. Yet there is a problem with this traditional work. Usually a broad-brush study will reveal an estimated return on investment that turns out to be very close to the return estimated after very complex and very detailed analysis. Refinement in the evaluation process hardly influences the return on investment, and this return is the key criterion upon which choices are made. Intuitively, mining companies and financiers understand that detailed studies are essential. The problem is that all of the extra effort and cost of these studies do not result *ex ante* in any reward—at least by the primary indicator of reward: return on investment.

For any given orebody and mining technology, the return on investment is often *less dependent* on any particular mine plan than on factors that are independent of the plan. The geological and metallurgical characteristics are largely independent of the mine plan. The selling price, exchange rate, and distance to market are largely independent of the mine plan.

Misguided detailed planning can of course *reduce* the return on investment quite easily. This aside, detailed analysis as traditionally practiced is aimed at *refining* the estimates within a well-defined set of constraints. Unless these constraints are reexamined or additional ore is found, the competitive position of most projects changes little through the evaluation process. This competitive position can be assessed after only a small amount of work (perhaps as little as 1 day of work) once preliminary ore characteristics are known.

This fact is quite well known and applies to many industries. If the return on investment does not change much with more refined planning, then two conclusions seem to follow:

- Early-stage evaluations are probably quite reliable indicators of the fundamental economics. If these early-stage evaluations do not demonstrate a satisfactory return, then technical priorities should probably be focused on continuing to look for other projects that do. This is the entrepreneurial, or exploration-focused, approach.
- For projects that do meet the investment criteria (after the small amount of initial work), there appears to be little value in any more detailed assessment when the expected return on investment will probably change very little.

The first of these conclusions is valid. The discovery of high-grade orebodies and completely new processes has been an important road to success for many companies. The second "conclusion" bears reexamination. Major mining companies do not commit to important new developments without a detailed study. Finance houses insist on detailed studies before they agree to debt funding. These organizations would not demand this work unless they were convinced of the value of it.

There is another disturbing trend evidenced through the way projects are designed and implemented. Superficially similar projects with similar mining methods and the same expected returns (initially, at the feasibility stage) are—after, say, 10 years of production—achieving vastly different results. Achieved results correlate only very poorly with expected results. Clearly some elements leading to actual results are not faithfully portrayed in the planning.

The evaluation process—feasibility study, discounted cash flow analysis, sensitivity analysis—examines the economics or competitiveness of the resource for a given set of conditions. The processes say little about the ease or difficulty in achieving the expected costs. They say even less about the ability of the project to sustain the returns under any other set of conditions—let alone how the project might work under some entirely different set of conditions.

The difference in performance (i.e., ability to live up to expectations) evidenced by the disparity in achieved results rests with the ease or difficulty in implementing change. This ease or difficulty may be an orebody characteristic or a market characteristic. Equally it might be a management characteristic—how efficiently decisions are arrived at and how correct these decisions are under the circumstances.

"Traditional" feasibility work is focused on study of a plan or variations of a plan to ever greater detail but within a relatively fixed set of rules and a relatively unquestioned external environment. The focus should instead be on studying the robustness of a project (perhaps to less precision) under different rules and under changing external environments.

A project that is robust under conditions of great change *should be* more deserving of acceptance despite a lower "expected" return on investment than a project that is not as robust.

SHORT-FORM EVALUATION TECHNIQUES

Once a mine is operational and comfortably fitted within a portfolio of mining assets, changes have to be made as the mining conditions and market requirements change. Continuing investment after initial mine establishment—frequently foreshadowed anyway in the feasibility study—is again subject to the decision-making process, as are mining alternatives that do not require new investment.

This defines one of the major difficulties in performance achievement in mining-style investments. Each "investment" is a decision to follow a certain path into the future, but the foreshadowed decisions along that path are not fixed. A rational choice foreshadowed at a time prior to the investment may be quite different from the choice that would be made once capital has been sunk. This applies even if original conditions are unchanged.

Original conditions seldom remain unchanged. A comprehensive DCF analysis cannot be undertaken each time they change. In this operating environment, the majority of decisions are made by using some short-form evaluation technique—a technique for which the recognized imprecision is accepted in return for the ease of calculation or implementation. Techniques are selected to suit the circumstances.

To reiterate the distinction argued earlier: The difference in performance (i.e., ability of a project as a whole to live up to expectations) rests with the ease or difficulty in implementing (ongoing) change.

Short-form evaluation techniques cannot be avoided. The difficulty is that, in choosing a technique, the decision maker is judging certain characteristics (those overlooked by the technique) to be unimportant. He or she is making this judgment based on experience, which will never exactly align with the task at hand or the circumstances at the time. Short-form techniques are generally reliable if the fundamental economics that underpin them remain true. This section reexamines the fundamental economics of some of the more common short-form decision-making tools.

The analysis subdivides the tools into two groups. The first group pertains mainly to operational or investment decisions in individual mines or certain types of mines. The second group is concerned with the mining business generally—the types of strategic choices made about whole portfolios of mining-related assets.

Operating Mines

Personnel involved in almost any production operation have little time for refined analysis. Refined analysis takes time, and the additional cost and lost revenue during this time frequently account for more value potentially lost than potential gains through optimized choices. This section examines some of these trade-offs and some of the common short-form techniques in operating mine environments.

Strip Ratio or Waste:Ore Ratio This first short-form rule applies only to open pit mining.

For any open pit mine, there is a strong correlation between the whole-mine economics and the amount of waste that has to be moved for each ton of ore or coal. Thus, this ratio—the strip ratio, or waste:ore ratio—is relied upon by many operators as an indicator of economic value. How reliable is this indicator?

In its simple form, the strip ratio number suggests that mine *costs* are a function of the amount of waste and that mine *revenues* are a function of amount of ore. A low ratio means low costs in relation to revenue and, therefore, is an indicator of relative profit. The reliability of the strip ratio as an indicator revolves around how true these representations are. Mine characteristics that upset this simple relationship are presented in Table 13.1. Some of these aspects have been discussed by Runge (1988).

One great advantage of the strip ratio calculation is that it can be performed in just a few minutes. More reliable methods (e.g., cost ranking) may take several days to calculate. The more reliable methods are warranted only if the value of the *extra* reliability exceeds the *extra* effort.

Two trends are evident:

1. As mining advances into increasingly complex orebodies, the simplistic strip ratio calculation becomes less and less reliable. Except in the simplest deposits, this tool is probably much less reliable than operators generally acknowledge.
2. The effort to calculate a reliable alternative measure of economic value is reducing daily because of advances in computer technology.

The economic value calculation is such an important guideline in all aspects of planning and operating a mine that it cannot be treated superficially in the interests of expediency. With the availability of alternative calculations, the burden of proof now rests with the strip ratio to justify its continued use in all but the simplest mines.

TABLE 13.1 Reliability of the strip ratio as an indicator of economic value

Characteristic	Comment and/or Adjustment
Low strip ratio	Low-ratio mines have a significant proportion of their cost associated with mining the ore or coal. In this case, the ratio of (waste + ore)/ore is better than the strip ratio as an indicator of relative economic value in an operating mine.
Mines with varying dip, varying hardness, etc.	For the strip ratio to be a faithful indicator, a volume of waste or ore in one part of the mine has to have a similar cost as the same volume of waste or ore elsewhere in the mine. If the dip (or other significant geological influence) varies throughout the mine, then costs *per unit quantity moved* vary from block to block.
Depth	Large, open pit metalliferous mines frequently remove all material from within the pit by using similar equipment (loaders and trucks), and the cost of mining is primarily a function of depth. Some operators adjust the strip ratio calculation for changing haulage cost with depth.
Equipment type	In many mines different equipment is used for different horizons and depths. Weathered material might be taken with one type of equipment and harder, deeper material with another. In coal mines, loaders and trucks often remove upper waste (prestripping) and draglines handle lower waste. For the strip ratio number to faithfully represent costs, complex adjustments are necessary. Low-cost dragline waste or truck-haulback waste in the bottom of the pit is low cost only because of the higher-cost advance stripping to facilitate it.
Operating cost or total cost?	Even if the strip ratio faithfully represents the relative cost/revenue, are the costs *total* costs or only operating costs? For short-term guidance, operating costs are the more faithful indicator of costs. For medium-term or longer-term guidance, total costs are more appropriate. Differences in capital-intensiveness can lead to differences in result.

Focus on Production: Incremental Production Many major resource projects are developed with large initial investments in relatively unproductive capital—fixed capital expenditure that is largely independent of production. Typical expenditures of this type include administration facilities, towns, roads, rail loops and whole railway lines, ports, and other infrastructure. The return on this capital must be covered by the initially planned mine output, but the return from incremental output has to cover only the incremental investment. In these circumstances, any expansion that yields increased production is usually quite profitable—even at marginal operating costs above the mine average.

In this environment, a culture has developed that suggests mine economics can "always" be improved by increased production, even to the exclusion of techniques that are more profitable but perhaps yield less or no production increases. How reliable is such a simple rule of thumb? If mine economics can "always" be improved, where does "expansion" stop? Can the whole deposit be mined out in 1 year? Or less?

If mine economics can be improved through expansion, then why wasn't the mine started at the higher production rate in the first place? What has changed between then and now? There are two strong reasons for expansion that are quite valid:

1. The orebody may have been only poorly understood at the time of original mine development yet was still economical. With extra understanding and extra reserves, mine expansion is rational.

2. The markets for mining products are often quite inelastic in the short term. Customers have large investments in unique capital, and increased offtake from new mining developments requires restructuring of the capital in these downstream processes—a time-consuming task. Thus, for these markets a steady expansion of production is consistent with this readjustment process and avoids the low marginal revenue difficulty shown in the example in Table 4.1 (p. 43).

If the short-form rule to expand is valid, the change should ideally be unique to the individual mine. If the impetus for change is also something that applies across a whole spectrum of industry, then expansion may be viable only as a defensive strategy.

Indeed, some of the best incremental investments in operating mines are investments that do not result in production increases. However, the problem with using discounted cash flow analysis for these types of investments is that for many people they demonstrate no identifiable source of revenue. (The "revenue" derives from the [savings in] operating costs that would otherwise be incurred. A savings in operating costs—particularly where good management is required to bring it about—is clearly not as evident as revenue derived from the sale of additional mine output.)

The difficulty in the expand-to-improve-mine-economics focus is that it frequently introduces a bias. Investment decisions are biased in favor of projects that result in expanded production over projects with equal or higher returns that result in lesser or no expansion but often constitute less risk.

Focus on Production: Avoiding Loss of Production Operating mines are judged primarily on their ability to meet production targets. Only a very few people on the mine site are concerned with *cost* per ton, and even then the criterion is commonly *operating* cost per ton. Operating costs *may* include depreciation provisions, but otherwise they overlook efficiency in the usage of capital.

This production focus is a reasonable criterion in most circumstances. A high proportion of capital is a genuinely sunk cost and can be correctly excluded from future decision-making influences. *Every* mine is most sensitive to loss of revenue. For any given selling price (usually outside the mine operator's control anyway), loss of revenue equates exactly with loss of production.

The problem with this singular focus is that *anything* that is likely to avoid interruption of production and hence disruption to the revenue stream is likely to be deemed justifiable. One dollar of prevention (advance planning) might achieve the same result as $1 million of remedial measures, and a singular focus on production does not provide sufficient incentive to balance the cost-effectiveness of prevention with the cost-effectiveness of remediation.

The archetypal case in mining involves pumping. A pump costing $100,000 is difficult to justify during the dry season. However, the lack of a pump causing the loss of a 50,000-t shipment can easily be demonstrated (albeit, too late) to be worth $100,000 on that shipment alone—and much more than $100,000 for the total number of lost shipments. Indeed, remedial measures (e.g., overtime work) to make up for the lack of the pump are commonly invoked at a cost far exceeding the cost of the forward-planning alternative (but at a cost still much less than the alternative of missing the shipments).

The loss of cash flow occasioned by the remedial measures is small compared to the alternative loss of cash flow that would apply with loss of production, but both are losses (in comparison to the long-term plan solution). Mining is a business of making profits, not of avoiding losses. The criteria for decision making in a "loss" situation should not be applied to longer-term investment aimed at profitability.

Measuring Production, not Production Cost The focus on production also extends to the guidelines used in day-to-day monitoring of mine performance. It is not economically feasible to monitor everything that is happening in the mine. All mines reach a compromise with respect to monitoring things that are important in terms of cost per ton, things that are important to production, and other things that are simply easy to measure.

There may be a complicated mechanism to translate measured data into knowledge about what to do. The example described in the Chapter 10 section entitled "Technical Analysis: Waste Removal" (p. 144) highlighted how the criteria for decision making often cannot be deduced even from a perfect monitoring system. What to do is commonly a function of the expected marginal cost, whereas measured data are always in terms of historical, average cost (often apportioned according to historical rules).

The clearest example has to do with loader productivity and truck productivity. Almost every mine using this equipment reports loader productivity daily. Very few mines report truck productivity. Because trucks travel to different places daily, there is no ready benchmark upon which to gauge their performance. Yet trucks commonly represent 70% of the cost of running a truck/loader fleet. Further, there is frequently an inverse correlation between truck productivity (not monitored) and shovel productivity (which is monitored).

If the criterion by which a production operation is judged is "production," and if the measure of production is *loader* production, then this target can be readily met by over-trucking the loader. In the example in Table 10.4 (p. 146), the addition of an extra truck from the optimum four-truck fleet to the suboptimum five-truck fleet increases production by 11% and makes a barely measurable 4% increase in (average) cost. Yet the collection of *average* costs statistics and the use of loader (or total fleet) productivity as the benchmark overlook the more correct *marginal* cost and *truck* productivity indicators. *Truck* productivity declines by 11%, and the *marginal* cost of earthmoving by the use of this truck in the fleet is 40% higher than the cost that would be incurred without it. Production (as measured) is actually an inverse proxy for the *real* objective of cost per ton. Measured costs are also a poor proxy for decisions, in this case suggesting the correct direction (increase or decrease) in cost following a decision but grossly underestimating the magnitude of the cost impact. In this example, "costs" as measured suggest a 4% increase in (average) cost if the fifth truck is assigned, when the real guidance for management decision making—i.e., the marginal cost of earthmoving using the fifth truck—is actually 40% higher than the (average) cost with the four-truck fleet (see Table 10.5, p. 147).

Focus on Capital When any new project is proposed, one of the first questions asked is: How much will it cost? There is a clear focus on capital. The focus is interpreted as a primary objective toward *minimization of capital.* This interpretation, introduced first in Chapter 3, also needs to be applied with caution.

The shareholders in a company invest with the express purpose and expectation that it is the business of the company to use their funds to make money. With capital constraints, the focus on capital should be directed at an *appropriate* amount of capital considering the company's funding structure, as well as other issues such as diversification and risk.

These issues notwithstanding, capital should be directed in order of projects yielding the highest return for the same degree of risk. If the company has opportunities that have an expected return of 25% or more, then projects only offering 20% returns will certainly be capital constrained.

Without a comprehensive examination of risk, return, and the cost of capital, there is a danger in using a rule of thumb that implies minimization of capital. Alternatives that are less capital-intensive (lower capital, higher operating cost) may be insufficiently robust under changing market conditions. Projects selected on this basis may be subject to much greater market risk—a situation perhaps quite contrary to the net objective of the organization.

Mining Strategy

Companies with whole portfolios of mining assets have to make decisions in the same way as managers on individual mine sites. Not all of these decisions can afford the time and cost of detailed analysis, and similar short-form rules are commonly used to guide such decisions.

In addition, the inputs to many decisions in a corporate-wide sense cannot be readily quantified, and senior management must again fall back on more generalized guidelines to aid choices. This section examines two such corporate-wide short-form guidelines: (1) the historical influence of inflation on choices and (2) the sustainability of returns deriving from a company's core competencies and its capability for organizational learning.

Culture of Inflation Operating mines and finance institutions associated with the mining industry have developed certain rules of thumb and other guidelines stemming from the culture of inflation since the early 1970s. Many of these rules of thumb were wrong even in this inflationary period. In times of low inflation or deflation, application of these rules can lead to gross errors in decision making. It is instructive to revisit the economics of inflation and reexamine the logic that underpinned the policies adopted by governments and investing firms during this period.

Inflation is primarily a monetary phenomenon. It is caused by increases in the money supply, and this money supply is largely under the control of the world's central banks.* Inflationary policies are pursued by governments to promote what they consider to be desirable outcomes. Inflationary policies typically aim to reduce interest rates because this promotes investment. If interest rates are lower–i.e., the cost of capital is lower–then, assuming nothing else changes, some mines that might not otherwise be viable become viable. In addition, for companies considering different methods to develop a mine, the methods that are more capital-intensive become relatively more attractive.

There are other implications–or at least *apparent* benefits of inflation. With inflation, a mine developed now will become more competitive over time because competing mines that start up in the future will cost more to develop. The apparent benefits also apply to employment. Wage increases usually do not keep up with inflation, so (assuming the selling price of the mine's product *is* escalating with inflation) the mine economics are improved because the real cost of labor has declined. When the real cost of labor declines, more people can be hired. Unemployment goes down.

These are the arguments that have been advanced (and still are in some places) favoring inflationary policies. Yet real-world experience has shown them to be quite in error. Not all the recognized errors have found their way into new rules of thumb for corporate decision making.

There is little debate among professional economists or policy makers that if the cost of capital declines, or if the price of labor declines and all other factors remain unchanged, investment becomes more viable. But do "all other factors" remain unchanged? If there is no inflation and someone has money in the bank earning 5% interest, will that person leave it in the bank earning 5% interest when there is 10% inflation? No. An investment in pure commodities (e.g., gold) that earns *no* interest but for which the value maintains pace with inflation is a more viable proposition. When the supply of loanable funds decreases, the cost of capital will rise, not decline. Indeed, "inflation" itself is not even necessary for this to happen. All that is necessary for investors to withdraw their funds from the loanable funds market is for them to *anticipate the likelihood of inflation.* To prevent the supply of funds from declining, interest rates have to rise–at least enough to counter the anticipated inflation. The cost of capital will *rise.*

What was once thought to be an effect of inflation is now recognized as a phenomenon of *unanticipated* inflation. The same is true of wages in an inflationary environment. Wages will lag inflation (and therefore represent a real cost reduction) only while the wage demands anticipate inflation at a lower rate

* The central banks cannot control the money supply precisely. In a fractional reserve system, the commercial banks expand and contract the money supply through their lending practices. In addition, there are many substitutes that have money-like characteristics (e.g., gold, precious metals, insurance policies, bearer instruments, and, of course, foreign currency). In the long term, however, the integrity of the currency rests with the policies that are pursued by governments, and inflation due to "wage pressure" or "price increases" is a misnomer–these are effects, not causes, of inflation.

than actually occurs. To be useful as an ongoing policy to promote investment, inflation must always be more than people anticipate it to be.* With the speed of modern communications, the availability of noninflating money substitutes, free exchange of currency, and the efficiency of world markets, the potential for governments to benefit from deliberate inflationary policies is very limited.

Although inflation may no longer be as significant a factor in investment choice, there are remnants of this previous high-inflation era that still influence decision making.

The first remnant is one of confidence. Much decision making is more a function of confidence than of value (in the net present value sense). This confidence or uncertainty precondition is explicitly recognized in the model in Chapter 12. Knowing that government itself is a major beneficiary of inflation and how politically unpopular some anti-inflationary policies are, most businesses were sure that inflation would continue. This trend was about as certain a trend as anyone in business could hope for. Yet the trend has been shattered. Paradoxically, the business environment in many parts of the world is now subject to more uncertainty awaiting government reaction to increasing international competitiveness.

The second remnant from the inflationary period is the trend of rising prices—particularly in the mineral industry. The retreat to real assets in the inflationary period was beneficial to the mineral industry and caused expansion of industries such as gold, silver, platinum, and (because of high oil prices) aluminum and coal. With low inflation this misallocation of resources is slowly unwinding, but the concept that prices generally rise (in real terms) remains. The idea is supported by the recognition that mines continually get deeper and lower grade, and the higher costs therefore translate into higher prices.

The reality is that real prices for almost all mineral commodities have declined, despite this trend to deeper and lower-grade ores. (The author has been unable to find an example of any major mineral commodity for which the price has risen in inflation-adjusted terms over an extended period. Some rising price trends *have* occurred in goods with boutique value or in mining products that have fallen into disfavor, but this trend has been accompanied by dramatic reductions in volume.) Real costs of production have also declined. Nevertheless, few mining companies have seriously adopted new guidelines for project evaluation using these trends. Most analysis continues to assume selling prices that rise according to inflation and costs of production that change similarly. A change toward *declining* selling prices and *declining* costs of production might yield substantially different emphasis on mining alternatives and management priorities.

---

* With ever-adjusting expectations of higher and higher inflation, the end result is hyperinflation as experienced in many countries this century. When inflation has to be unwound (by reducing the money supply), a reverse problem occurs. If people do not believe that the government has the resolve to address the issue, then anticipated inflation exceeds the real monetary inflation, causing wages and prices to remain higher than they would otherwise be—a recessionary scenario. This was a major issue in the United States in the early 1980s when inflation declined dramatically.

Core Competency and Models of Learning The short-form evaluation techniques described in the "Operating Mines" section earlier in this chapter are useful in circumstances where a more comprehensive approach is unwarranted. If there is ever any doubt about the reliability of the technique, more comprehensive analysis can be undertaken and the answer determined with reliability.

For strategic choices concerning whole portfolios of mining assets, there *is* no analysis that can guarantee reliability. Even if a comprehensive DCF model could be built up of company A's mines and all of its competitor's mines, this will not necessarily tell company A what the competitor is likely to do and therefore what company A should do. Suppose, for example, that company A's assessment correctly showed a potential return of 20% for a proposed new project and a potential return of 15% for the competitor on its new project. If only one new project is viable in the market, does this mean that the competitor, *even if it had this information,* would not start? No. The difficulty is that company A cannot know the competitor's (opportunity) cost. If the next most attractive alternative yields the competitor a 10% return, then it is viable for the competitor to commence, even though its project is notionally inferior to company A's project.

How, then, can whole portfolios of mining assets be viewed and managed in such a way that their economic value can be sustained over their life?

Very little work on this issue has derived from mining applications, but extensive analysis has been undertaken in more general industrial organizations, and this work is increasingly finding its way into mining. This section examines some aspects of this transition.

The easiest way that long-term competitive advantage can be sustained is via some unique input to the process. The formula for Coca-Cola® is the quintessential example. Patents, copyrights, and trademarks are also unique inputs. Mining companies always have one unique input—the orebody. The difficulty with this approach is twofold:

1. Unique inputs lose value over time. Orebodies that once were rich and shallow become deep and low grade. New technologies evolve to make old patents worth less. Tastes change. A long-term strategy is not built on the *holding* of unique inputs, but on the ability to continually *discover* new unique inputs or enhance the value of existing ones. In the mining industry, this ability is a characteristic of an exploration-focused company, not a mining company. This valuable entrepreneurial skill should not be confused with the skills needed for the *operation* of mines.
2. Newly discovered orebodies, patents, and copyrights can be sold or licensed. If these inputs indeed underpin long-term profitability, the present value of them—the extra value they add to any process—can be captured by the discoverer from the start. This happens on a regular basis when junior exploration companies (whose skill is finding such orebodies) sell off part or all of their interest in their newly discovered orebody to a mining enterprise. If the ownership passes to the highest bidder, then

> there is no "surplus" profit to underpin long-term *operational* competitiveness. Mining enterprises achieve operational competitiveness through some skill in *operation,* not through finding the orebody in the first place.

If an enterprise is to sustain itself for the long term, then unique inputs are still required—but they must be inputs that are a feature of the organization itself. These inputs cannot be sold except by selling the organization. Further, not only must the organization possess unique inputs, but mechanisms must also exist for updating, expanding, and extending these unique characteristics. In the terminology of Prahalad and Hamel (1990), these unique inputs are commonly referred to as core competencies.

Mechanisms for updating, expanding, and extending knowledge can be referred to as *organizational learning*. If an organization produces the same item year in and year out, and if the culture of the organization encourages it, then better and better ways will be found to produce the item in a faster or cheaper manner. Even if another organization were to purchase the identical plant and machines, there is no guarantee that they could replicate the institutional knowledge that provides the efficiency to the incumbent.

The competitive advantage from sustained organizational learning has been recognized at least since the mid-1940s. Early work most commonly cites the production of aircraft. For example, Alchian (1963)* studied the number of direct person-hours needed to produce each airframe and developed a relationship of how this number declined with the cumulative number of airframes built. This study and other studies have shown no practical limit to cost reduction. The more experience an organization obtains (the more *total* production), the more the number of inputs (in this case, direct person-hours) continues to decrease.

The implications of this have been recognized in a number of short-from business models. (The most notable model in this category is the "Growth-Share" matrix developed by the Boston Consulting Group. Rothschild [1992] includes a discussion of this model and other examples drawn from research into learning curves similar to Alchian [1963].) Assuming a competitive market where everyone sells at the same price, the most profitable company will be the one with the lowest overall cost of production. If there are no other unique inputs and all organizations have the same capacity to learn from experience, then the lowest-cost producer will be the one with the most experience. Thus, market share becomes an important strategic aim. Organizations with high market share advance along the experience line faster than organizations with low market share.

Many large mining companies now use these simple models to guide decision making across their portfolios. How reliable are these models, drawn from industrial organizations, likely to be for a mining enterprise?

---

* Original work undertaken in the 1940s by Alchian was "military classified" and made available only in the 1963 paper.

The key ingredients to success using this approach are as follows:

- The models focus on the difference between the selling price and the cash cost. Sustained competitive advantage rests on the difference between selling price and cost compared to competitors in the same market. Thus, some companies prefer to focus *directly* on this attribute—aiming for projects in the lowest 25% in terms of costs of production.
- The models assume competitive markets where everyone sells at the same price. However, not everyone sells at the same price. A high-cost producer can still enjoy a large difference between selling price and cost if output can be sold at a higher price. Markets that are developed to capitalize on unique characteristics of particular ores can often achieve this differentiation. The mineral product does not require superior characteristics to be differentiated; rather, its characteristics have to be unique *only in the mind of the customer.*
- Market share should not be confused with cumulative experience. Some organizations pursue market share as if it alone will deliver efficiency. A new company "buying" market share (through deliberately underpricing) cannot automatically reduce its real costs of production any faster than a long-established organization with more experience continuing to supply a lesser percentage of the market. Mines scheduled for production rates in excess of some natural rate for the deposit suffer inefficiencies from "too much equipment trying to fit in too small a hole."
- Cumulative experience should not be confused with learning. Cumulative experience alone cannot deliver efficiency. Efficiency comes from how much an organization *learns* from this experience. Organizations have different propensities to develop and retain skills. The reduction in costs in the short term (through retrenchment of noncore people, for example) can often dramatically reduce an organization's capacity for learning and continuous improvement. (Senge [1990a] develops this model of learning as a key ingredient in sustaining competitive advantage.)
- Learning is not restricted to refinement of mechanical skills. It is easy to see how repetitive tasks in manufacturing can be progressively refined over time. Mining rarely involves repetitive tasks of this nature. Orebodies are mined out and change with time. Mining skills are apparent less in the tasks themselves than in the decision-making processes that allow similar *types* of tasks to be repetitively undertaken. Love (1997) suggests that the core competency of the mining corporation is (strategic) decision making.

Strategic decision making requires more than just a production focus. Most economic models that examine strategic decisions in a competitive environment start with an assumed competitive selling price and a strategy that has producers meeting this price at lower cost than competitors. In a consumer market, this is a reasonable starting point for sustained advantage. However, the products of mining are seldom sold in the consumer market. Mining is typically just the first step in a long chain of production, and its products are quite remote from the end user. Changes in the structure of production can dramatically change the nature of the market. Direct reduction of ores can bypass traditional multistep processes. End products can be made out of different materials. How should the market be defined? Is it a market for "iron ore" or "aluminum," or is it a market for "the structural elements used in motor cars"? A narrow focus on market share or operating cost, in an industry positioned so early in the production chain, risks dramatic losses in value through changes in this structure of production.

Some of these issues are taken up again in Chapter 16.

CHAPTER 14

Decisions Involving Uncertainty, Risk, and Return

This chapter looks at discount rates and the relationship between risk and return. It develops the conceptual tools for making comparisons. It sets out two case studies that illustrate how to value uncertainty in an analytical way. It sets out guidelines for applying these trade-offs in a generalized, objective way.

DETERMINING THE APPROPRIATE DISCOUNT RATE

The valuation of capital assets is a function of the expected future earnings (including the probability that they will materialize) *discounted* at a rate that reflects the time and risk preferences of the participants.

Low discount rates reflect more confidence in the future and increase the present value of all projects. Under low discount rates, projects with a longer-term payoff become *more* viable relative to projects with shorter-term payoffs. Nevertheless, low interest rates (and low discount rates) are not necessarily good or better than high rates. Developing for the long term is not necessarily good or better than focusing on the shorter term. The highest rewards will go to the business enterprises whose products and services are most consistent with demands in the economy as a whole.

This section is usefully started by stating up-front one of the primary problems with discount rate determination as commonly applied: that companies seek a certain minimum return under a strong expectation that it will not be forthcoming. Most companies might be happy with, say, a 10% return on investment if that return had a strong guarantee of materializing. Yet because of poor past performance or an inability to define investment requirements in a more robust way, they set their acceptance or rejection criterion at, say, 20%. In so doing, they tilt the economic playing field toward certain types of investments to the exclusion of other investments that offer greater potential to achieve the (real) objective. This is not the path to more reliable decision making.

The criterion to be used by a *company* may be quite different than that used by individual shareholders. Individual shareholders engaging in risky investments on their own account will limit their total investment to a relatively low figure. A company whose shareholders all have the same characteristic aversion to risk can happily invest many times this amount because of diversification. A project likely to pay off handsomely after 15 years (but pay nothing prior to this) may be of limited interest to *individuals* yet may still be appropriate for a company made up of the same kinds of individuals. Earlier returns are realizable by individuals through stock appreciation, even if the project itself has not yielded any returns. The path from individual risk/return trade-offs (in the mind of an individual shareholder or individual board member) to corporate-wide risk/return trade-offs is interspersed with many irrelevant sidetracks—all of which inhibit fulfillment of the objective.

Mining industry investment traditionally demands a higher return because of the higher perceived risk. Nevertheless, simply adopting a higher threshold rate does not resolve the issue in cases where alternatives are subject to *different* risk. For sustainability of long-term returns, the threshold expected return on investment for a particular project should be greater than or equal to the cost of capital for that project plus premiums for

- the cost of exploring for and evaluating new projects to ultimately replace the economic reserves being depleted by this project
- the cost of maintaining the company "knowledge" base and other intangible company assets to actually deliver operational capability on this and any (future) replacement projects
- additional risk associated with the project until it starts performing with sufficient consistency so that it can be assessed by the marketplace

Unlike with the manufacturing industry, the rate at which existing knowledge is made redundant is much higher in mining. As mines advance, mining techniques change. As a result, the mastery that the company has acquired for the existing technique is no longer relevant. Some of these knowledge problems are addressed again in Chapter 16.

The above criteria are sufficient only for *sustainability* of returns. A company will grow more quickly insofar as returns exceed these starting requirements. Growing earnings (when discounted to the present) mean the company in total is worth more. An *expectation* of growth will therefore result in a reduced cost of capital through share price appreciation—at least while this expectation is sustained.

Historically, the average return from mining company portfolio investments has not been any greater than for any other grouped investments in the stock market, after adjustment for higher volatility compared to the market as a whole. This is to be expected. Unless a company is purposely harvesting its returns from existing projects, its performance will of necessity be reduced by the expenses of

- absorbing ongoing exploration
- absorbing and assimilating technology for the next phases of developments without being able to fully capitalize on the "just-learned" technologies

- absorbing the losses from the proportions of projects that yielded the lower-than-expected return, insofar as they were not counterbalanced by the projects that yielded the higher-than-expected return

There is a big difference between expected returns in the stock market and expected returns from individual companies and projects. Individual projects do not have sufficient diversity and must be assessed on a case-by-case basis. Nevertheless, even for one company with *all* of its investments performing according to its investment criteria, the costs of renewal (finding replacement deposits, upgrading technology) will always reduce the return to the company to a number less than individual project return.

EXPECTED RETURNS AND THE COST OF CAPITAL

The expected return that a new project must yield is ultimately determined by the return available from the best alternative use of a company's resources. The number of alternatives available to a company at any one time is limited, and most firms use guidelines for their "required" return. New projects must demonstrate a return exceeding this guideline to be considered, even though actual approval will still be subject to prioritization of capital resources according to return.

If the world were a fully competitive and economically efficient place, then there should be very little difference between the return on investment being sought by companies, on the one hand, and the cost of capital on the other. Yet in practice very large differences are observed. Summers (1987a; 1987b) surveyed 200 large corporations, aiming to discern the applied discount rates and how they varied for elements of the cash flow that had different risk characteristics. For example, depreciation allowances contribute to tax savings, and even if individual projects are risky the translation of these allowances into corporate-wide tax savings is subject to very little risk. The reported discount rates were "surprisingly high" (Summers 1987a, p. 32) with a median of 15% and a mean of 17%. (The rates in inflation-adjusted terms were not reported, but, given inflation rates during this period of 4–5%, real expected returns would be in the order of 11–14%). Pindyck and Solimano (1993), drawing on this and other studies, have also noted that "hurdle rates that firms require for expected returns on projects are typically three or four times the cost of capital."

These results are consistent with observed practice in the mining industry. A very common guideline used by large companies in the mineral industry is an after-tax discounted cash flow return of 15%, assessed on a constant money basis (over and above inflation). Most large mining companies enjoy an after-tax cost of capital less than half of this guideline.

Since this is a common guideline, yet few companies achieve this return on average, a reconciliation of the aimed-for return and the achieved return is warranted.

If a company is to stay in business in the long term, then the minimum return must at least cover the cost of capital. There are two primary reasons for the observed (large) difference between the "expected" return and the cost of capital. The cost of capital for an individual investment is not exactly the same as the average cost of capital for a firm. Moreover, firms are seldom risk neutral with respect to large capital investments, and their tolerance for risk influences the premium sought above the cost of capital. These two issues are addressed in the following two subsections.

Marginal Cost of Capital for Individual Projects

In Chapter 12 a model was prepared characterizing risk and uncertainty (see Figure 12.1, p. 183). In this model, outcomes that yielded a return less than the cost of capital were said to represent risk. Again, earlier in the present chapter, the starting point for selecting the discount rate was the cost of capital. This section examines what the "cost of capital" means in this context.

The weighted-average cost of capital for a mining company can be calculated relatively easily by using formulas in most corporate finance texts (e.g., see Brealey and Myers [2003]). If a large mining company based in the United States wants to develop a new mine in a well-understood mining region within a mining-friendly state, then this plan is easy for the market to assimilate. Additional equity or debt could be raised, and the cost of capital for this project (the *marginal* cost of capital) would be little different than the average cost of capital for the firm as a whole.

If the same company, not previously operating outside the United States, were unexpectedly to announce a major new development in an environmentally sensitive region in a third world country, then the equity and debt markets would react in a far from predictable way. The share price might fall dramatically. The marginal cost of capital using this indicator is substantially different than the long-term average cost of capital.

Even if short-term market reaction or individual project finance costs are not used as the indicators of the cost of capital, *long-term* debt and equity costs are still predicated on a historical expectation of repayment. If this expectation is called into question, then the cost of capital will change.

In an assessment of an individual project, what is the marginal cost of capital that should be used as the benchmark for determining risk? This is not a trivial task, and an example from Gentry and O'Neil (1984) illustrates how easy it is to use the incorrect values. The example (under the heading "Error No. 2. Using Specific Capital Costs as the Discount Rate" in the original text) is paraphrased here and then commented on.

This example cautions analysts *not* to use the cost of a specific source of financing but rather to use a weighted average of all capital sources as the discount rate in measuring the attractiveness of a project. Gentry and O'Neil draw upon an example from Quirin (1967):

Global Mineral Ventures, Inc. (a hypothetical company name) was presented with a similar set of investment opportunities in three successive years as shown in Table 14.1.

TABLE 14.1 Investment opportunities in three successive years

	Year 1		Year 2		Year 3	
Project	Amount of Investment ($)	Rate of Return (%)	Amount of Investment ($)	Rate of Return (%)	Amount of Investment ($)	Rate of Return (%)
A	100,000	20	100,000	20	100,000	20
B	200,000	15	200,000	15	200,000	15
C	200,000	11	200,000	11	200,000	11
D	200,000	8	200,000	8	200,000	8
E	200,000	6	200,000	6	200,000	6

In year 1, Global had no long-term debt and the corporate treasurer found that the full $900,000 for all five projects could be raised by selling debentures bearing an annual interest rate of 5.5%. He convinced Global's board that, since every project returned more than 5.5%, they should all be accepted.

In year 2, the treasurer was able to borrow a further $700,000 at 7.5% and, by using the same logic as the previous year, accepted all projects except E, which offered a return below the 7.5% marginal cost of debt.

In the third year, however, Global's treasurer found only a limited amount of additional debt available to him–$100,000 at 18% from a finance company. Since this was still below the estimated 19% for a new equity issue, the treasurer used the debt to accept only project A.

"Thus," suggest Gentry and O'Neil, "over a three-year period by using the cost of a specific source of debt as his investment criterion, the treasurer had invested $1.7 million at a weighted average return of 12%. It is clear, however, that the treasurer's eagerness to accept projects in year 1 precluded the acceptance of better projects in year 3" (p. 342). The same $1.7 million could have been invested to yield an average rate of 13.3% by accepting the following projects:

Year 1	A, B, C, D
Year 2	A, B, C
Year 3	A, B, C

What went wrong? Where was the treasurer's concept of marginal cost of capital in error? Gentry and O'Neil provide the first explanation:

> *In essence, the treasurer failed to recognize that debt financing is only possible if an adequate equity base exists. If expansion of the equity base does not keep pace with borrowings, the firm will reach a point–as Global did in year 3–where the firm's financial risk is too high for further credit at reasonable rates. Clearly the marginal cost of capital in this case, then, was not simply the cost of debt but also the cost of equity which will be required in the future to support the added debt. As Quirin notes, "When the capital structure is changed by an issue of debt, the relevant cost is not only the out-of-pocket cost of the debt itself, but includes the increase in the cost of equity resulting from the higher risk premium attached to the shares as a result of the debt." Thus, every investment must carry*

its proportionate share of the necessary–but higher cost–equity funds. In the above example, using the marginal weighted average of all capital sources would have resulted in rejecting the lower-value projects in year one, thereby permitting the acceptance of higher-value projects in year three. (pp. 341–343)

This explanation is only partly correct. Indeed, in criticizing the marginal cost of capital from this perspective, the explanation makes two additional errors itself, one strategic and one due to a confusion between a financing view of cost and the economic concept of cost.

In terms of making a choice today that makes choices in the future more expensive, this example has similarities with the example in the Chapter 6 section entitled "Impact of Production Rate" (p. 87). In the example from that section, the comparison was between something that a company *planned* to do today with something that it simultaneously *planned* to do in the future. In the current example, the treasurer in year 1 did not *plan* to invest in any projects in years 2 and 3. This possibility apparently did not occur to him. Outside observers have the benefit of hindsight. Gentry and O'Neil, however, expect the treasurer to have perfect foresight, and only under this scenario is the criticism valid.

There may be a valid criticism from a strategic view. The treasurer should have been aware of more possibilities than just the five projects offered to him. He might have asked: "What is different about this year compared to last year that makes these series of projects viable? Is the same thing likely to happen next year? If I am presented with a similar series of projects next year, how will this influence my choices for this year?" He might also have questioned his company's starting position with no long-term debt. Buying back his own shares with debt may have been a more attractive investment than project E in year 1.

The second error of this analysis follows from the first. Recall the definition of cost from Chapter 4. The "cost" of anything is the highest-valued opportunity necessarily forsaken. In this case, the financing "cost" as developed in the example is *not* the cost. The high-return projects in years 2 and 3 (that have to be passed over) are the cost. If the treasurer had recognized that these projects would have to be forsaken, then *despite the availability of finance* he would not have invested in project E in year 1.

The scenario is also incomplete from an economic, opportunity-cost perspective. In year 3, the treasurer is rejecting projects that yield 15% return but continuing to hold projects that yield 6% return. The return on these projects based on the *sale price* of the project may be higher than 6%, but nevertheless one or more of them may be saleable at a cost (loss of opportunity) less than the 18% financing cost. Thus, the cost of capital in year 3 may again be improperly portrayed by using the cost of *debt* finance. Higher-return projects in year 3 may still be possible.

The errors that the treasurer has made are errors of foresight and an inability to envisage a larger range of alternatives for action. The error by Gentry and

O'Neil is in equating the financing cost with the opportunity cost for decision making. There is no error in choosing projects on the basis of marginal cost.

Under the economic definition of cost, returns are maximized and investments should correctly be selected to the point where the marginal return equates to the marginal cost of capital. The definition of marginal cost for individual investment decision making does not end at this point. If there were no other opportunities available than the ones presented, does this mean that *project-specific* costs of capital should be used?

Again, caution is demanded. Lenders generally supply funding against a cash flow stream coming from diverse sources. Market rates on offer may not reflect any informed opinion by lenders of the likelihood of payment from specific projects alone. It is the "informed opinion" that is the key.

The correct marginal cost of capital from a source-of-finance perspective derives from how a *well-informed* debt and equity market views the project. Clearly, at the start, outsiders have no view at all. Indeed, the *best* opportunities are available only when no one else realizes they exist! Thus, the marginal cost of capital has to be estimated. It is the owner's *expectation* of how the project will be judged by the market once it is performing and information is available for outside assessment.

Thus, a point made earlier in this chapter—that the threshold expected return on investment for a particular project should include a premium for the additional risk a project bears until its performance can be realistically assessed—has to be recognized. Allowance has to be made for a market-adjustment process and the time that this process takes. This is the time from when a project is committed to the time when it is operating in such a way that outsiders *do* have an informed opinion about performance. Management is "on notice," or "at risk," until the stock market generally believes that what is planned will actually happen. Until this time, actual changes in share price may be only poor indicators of changes in the marginal costs of capital. Easily assessed investments require a lower premium. Complex, hard-to-assess investments demand additional premiums. Chapter 15 explicitly looks at this at-risk component of major capital investment choice as a further indicator of value.

Risk Tolerance: The Uncertainty Criterion

Once a project offers a return exceeding this cost of capital, can it be started? A large company investing in hundreds of small projects *can* probably proceed on this basis. New mining investments are not normally started by the hundreds, however, and the diversity that benefits small projects cannot be counted upon.

This lack of diversity foreshadows the second element in the investment decision process: a recognition that no rational investor participates in a major project with only a 50% chance of success. This is essentially what the "expected" return denotes. Clearly, to be considered, projects must have a high chance (perhaps a 90% probability) of exceeding the cost of capital. This "90% probability" criterion is the uncertainty precondition first outlined in Chapter 12. If there is a 10% probability of the return being *less* than the cost of capital, then the "expected" return (corresponding to a real return that typically has an equal chance of being either greater or lesser) must be substantially more than the cost of capital.

The uncertainty criterion is a general term for the risk tolerance of the organization.* In simple terms, if a firm wants new investments to have a 90% chance of exceeding the cost of capital, then they are more risk tolerant than a firm that wants a 95% chance.

There is also a second way of considering this criterion; this way can be expressed in terms of cost of capital, selling price, or any other input to the decision process. It may be expressed in question form as follows: Given current selling prices, what would the *cost of capital* have to be for the project to satisfy the tolerance for risk? The problem can also be framed in terms of selling price (or any other unknown subject to uncertainty): Given the cost of capital, what would the *selling price* have to be for the project to satisfy the tolerance for risk? It is this second way of framing the problem that opens up major possibilities for capital investment choice under uncertainty. The example given in the section "A Case Study: New or Old Equipment" (later in this chapter) illustrates.

The uncertainty criterion also applies in a more subjective way to choices that cannot be readily quantified. In these cases, the cost-of-capital measure is replaced by some reservation value that must be exceeded with high probability for a choice to be acted upon. Thus, the impact of local or global wars, political uncertainties, and industrial and environmental disruption can be assessed first via this precondition, at least in a subjective way.

Do firms actually make investments according to this model, or are firms normally risk neutral as suggested in most economics texts? *Individual* decision makers within firms can be risk averse while still making decisions on behalf of the firm in a risk-neutral way. Most observers would not support the risk-neutrality model in the minerals industry. For example, increased fluctuations in the selling price of a mineral typically cause a hiatus in new investment relating to that mineral even where the overall trend is for an increasing price. A risk-neutral firm would be influenced only by the overall price trend, not the fluctuations about this trend line.

Actual evidence (as opposed to anecdotal evidence) either for or against this model of investment behavior is difficult to establish. The difficulty is similar to the one that led to serious misunderstanding of the effect of inflation as described in Chapter 13. In this example, what was thought to be an effect of inflation was actually found to be a phenomenon of unanticipated inflation. If mining companies live with uncertainty year in and year out, then decisions will be made with similar regularity as they are made in other industries with

* Strictly speaking, "organizations" do not have risk tolerance; only individuals have aversion or relative liking toward risk. Risk tolerance in this context refers to the institutional processes within organizations that facilitate pooled understanding and residual responsibility for collectively-arrived-at decisions involving risk. These institutional arrangements vary enormously across organizations. Economic models that overlook these institutional arrangements (i.e., *most* economic models) assume that firms are risk-neutral.

lower levels of uncertainty. It is only when uncertainty changes (the future becomes less or more predictable than it previously was) that any changes in investment behavior are likely to become evident. Ferderer (1993) has attempted to develop such a forward-looking indicator with inconclusive results. Nevertheless, as Leahy and Whited (1995) point out, the inconclusiveness may be due to insufficient differentiation between changes in demand and changes in uncertainty about demand. In a study similar to mineral investment applications, Hurn and Wright (1994) have modeled decision making for oil field developments in the North Sea. The results of this study indicate that the variance in the price of oil did not impact the delay in oil-related investment, but again, this study had limited ability to separate out changes in price from changes in uncertainty about price. (For other work attempting to demonstrate the link between investment behavior and uncertainty, see, for example, Abel [1983], Abel et al. [1995], Abel and Eberly [1995], Lucas and Prescott [1971], Morrison [1993], and Runge [1990]).

Guidelines

While the difficulty in quantifying the model and many of the preceding issues is acknowledged, a number of adjustments to any fixed guidelines are self-evident.

1. Projects that enhance the *existing* businesses of a company have a lower cost of capital than projects that take the company into new areas. Markets will assign the lowest premium and the lowest cost of capital to businesses that they understand the best. For two projects exhibiting the same expected return on investment, the one that is most consistent with existing business is favored.
2. Projects that have extensive reserves—beyond the typical 15-year life of discounted cash flow studies—can proceed with lower returns because they have elements of value not normally included in DCF calculations. They contain reserve extensions to replace ore that is worked out. Further, the projects include an option to expand or extend, and this option is worth something.
3. The likely changes in technology throughout a project life should be considered. Projects for which the viability is sheltered from the pressure of change (e.g., coal mines supplying captive power stations) may proceed with a lower return because they demand a lesser commitment to funding corporate technology development.
4. Asymmetry of returns frequently stems from a project's inability to capitalize on beneficial events (e.g., short-term price increases) and the inefficiency of being reactive rather than proactive to adverse events. Projects for which the deposit characteristics, industrial climate, or financing and management structure limit this sort of change should indeed be discounted (i.e., require higher rates of return) in comparison with projects that have the scope to capture "upside" potential.
5. On the basis of the uncertainty criterion (probability of exceeding the cost of capital), projects with lower risks should be treated differently than projects with higher risks.

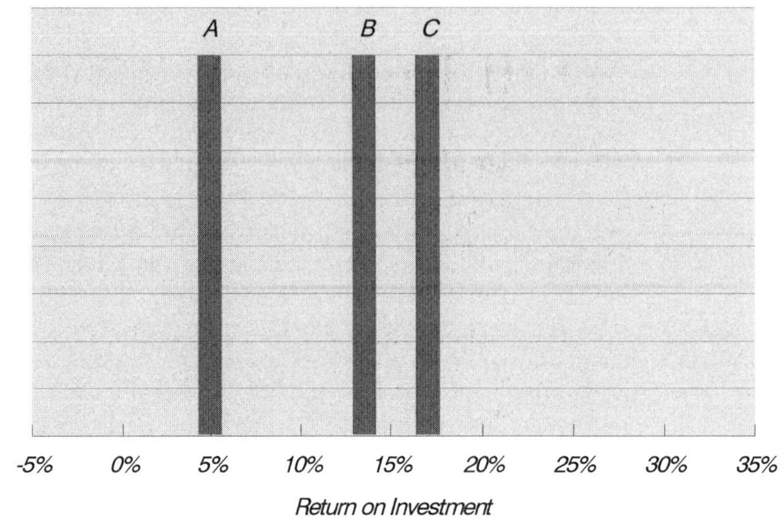

FIGURE 14.1 Typical presentation of return on investment

These points are worth elaboration. Consider a type of capital investment aimed at improving an existing business. Capital improvements frequently involve no additional output–the additional revenue to repay the capital comes from the reduction in operating costs. Contrast this with capital investments used to expand production. Capital *improvements* are easier to understand (point 1 from the preceding list), do not deplete reserves (point 2), and often reduce the sensitivity to technology changes (point 3). More important, the additional revenue funding them is *not* subject to market price risk (point 5). A substantially lower *expected* return for this type of project will satisfy the uncertainty criterion compared to the return required for a new mine investment yielding additional output.

CHOICE AMONG PROJECTS WITH DIFFERENT RISK

Within any one company, how can different projects be considered with different investment criteria depending on their degree of risk? What exactly *are* the risks, and what relative premiums should be applied for the asymmetries, better knowledge, and technological change factors outlined earlier?

Figure 14.1 shows the typical understanding of three alternatives for a project after evaluation using DCF techniques. Alternative A has a lower capital cost but a lower return. There is no vertical axis label in Figure 14.1 because the return-on-investment number is typically considered a monodimensional ranking-ordering number. If the full opportunity set is included in the analysis (i.e., "do nothing" and "delay the project" alternatives are included), even the magnitude of the return-on-investment number is of little significance–the highest-ranked return on investment signifies the best choice.

In practice the opportunity set of choices never includes all possibilities, so the magnitude of the (expected) return on investment does have significance.

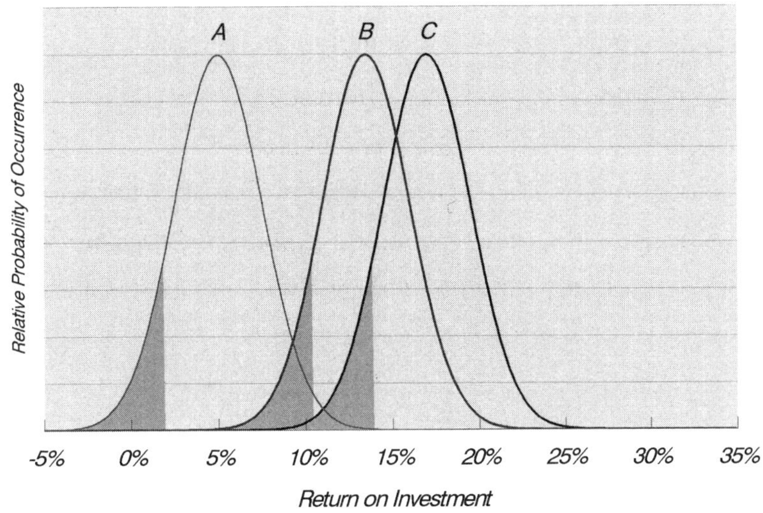

FIGURE 14.2 Project return on investment: probabilistic representation

If a 12.5% return is the primary criterion for investment, then alternative A is eliminated. Similarly, most companies implicitly if not explicitly recognize the higher marginal cost of capital for alternatives that require more capital. If alternative C has higher capital costs than alternative B, then the higher return from this alternative is weighed against the higher capital requirements.

The shortcoming in this assessment is the narrowness of the focus. Cases are presented as best estimates. In this case, *best* in the mind of the analyst usually means the most likely outcome. Yet in any presentation of best estimates, there are many things that might reduce the achieved return, and there are also things that will improve the return. Insofar as these outcomes can be envisaged, there is probably about a 50% chance that the return will be exceeded and a 50% chance that the achieved return will be less than the estimate. Figure 14.2 shows the same situation as in Figure 14.1 in probabilistic terms, assuming in this first instance that these distributions of outcomes are normal distributions.

The distributions have been drawn with the same variance, implying the same degree of uncertainty across choices. For the uncertainty criterion (90% probability) to be satisfied for project A, the cost of capital would have to be 1.8% or less. For projects B and C, the cost of capital would have to be no more than 10.3% and 13.8%, respectively.

Figure 14.2 may be a more faithful representation of uncertainty, but for projects with similar uncertainty it does little to aid decision making. Unless there is a difference in the cost of capital across choices, projects with similar uncertainty are ranked similarly, regardless of whether the rule for ranking them is net present value or the uncertainty criterion.

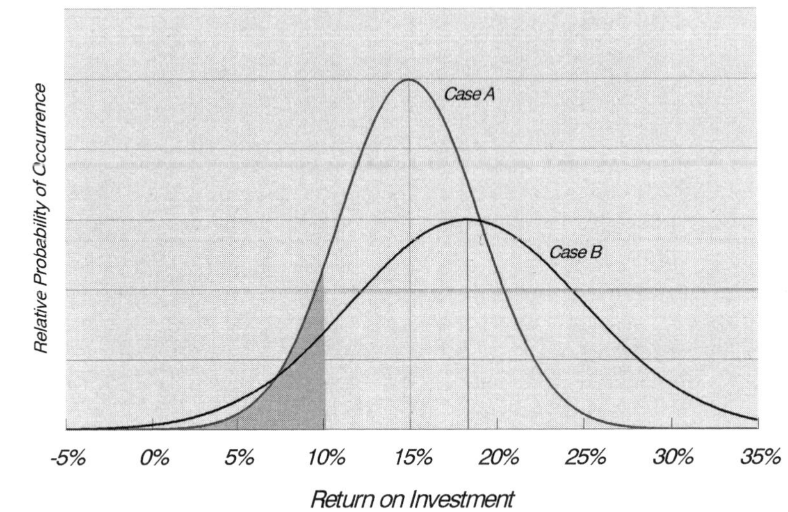

FIGURE 14.3 Comparison of projects with different uncertainty

Thus, companies that make whole-project comparisons primarily based on the internal rate of return or NPV measure—and (apparently) not explicitly considering uncertainty—are not necessarily committing any error. Differences appear only if the uncertainty is different across choices or the cost of capital is different across choices. For alternative ways to develop a mine, the major uncertainty commonly derives from variability in market price, and this variability is probably common to all alternatives. Similarly, the cost of capital is probably common to all alternatives.*

For projects where the uncertainty is different across choices or the cost of capital is different across choices, satisfaction of the uncertainty criterion (in an expectational sense) is a precursor to evaluation leading to choice on the basis of NPV (or IRR). Projects that have an expected return exceeding the cost of capital should be examined (drilled out, studied, etc.) until they are understood to the point where they satisfy the uncertainty criterion and then selected on the basis of highest net present value. Figure 14.3 shows this type of comparison between two projects each having the same cost of capital. In this instance, case B is not understood to the same degree of reliability as case A, but it *is* understood to the point where the uncertainty criterion is satisfied.

There is anecdotal support demonstrating that mining companies *do* evaluate projects in this way, even if they do not explicitly acknowledge it. New mines

* Runge (2000) examines models of choice under uncertainty and reconciles observed behavior with these models. The difficulty is that, where uncertainty is the same across choices, results of observed action are identical whether the actor is choosing on the basis of the uncertainty criterion or on the basis of NPV or IRR. Conceptually falsifiable results to this hypothesis are possible only where uncertainty is different across choices, and empirical work to date has not separated such choices in any analytical way. Nevertheless, the large number of firms found in some studies to use payback as an indicator of value suggests support for the model. As shown in Chapter 15, payback is a partial though imprecise indicator for capital "at risk."

TABLE 14.2 Sequential steps in decision making for large investments

Priority	Activity
1	All investment is "relative" to something (the "do nothing" case, usually). Clearly establish the likely economics of this benchmark. Scrutinize all possibilities, including share buybacks, reallocation of resources from existing projects, and possible future projects for which the viability might be influenced by the immediate decision.
2	Establish the broad economics of the alternatives. This establishes the expected return but leaves many uncertainty characteristics unknown.
3	Determine whether the project has an expected return exceeding the likely marginal cost of capital. An expected return exceeding the cost of capital is a precondition to investment *in information* aimed at risk minimization.
4	Invest in information (more evaluation). Ensure that both the (new) project and the most likely alternative that will be foregone are understood to a degree of reliability that satisfies the uncertainty criterion (less than 10% probability that the return will be less than the cost of capital). Continue the examination of alternatives (even alternatives that reduce the *expected* return on investment) until this criterion is satisfied.
5	For all projects satisfying this criterion, rank in order of (expected) net present value. Ensure that comparative rankings include everything that differentiates the projects (including intermediate cash flows). Consider logical projects as well as physical projects. Invest in projects in order of highest net present value.

that are fundamentally very profitable (high expected rates of return) are planned in less detail than mines that are less profitable. Underground mines that have a larger inherent uncertainty than open pit mines run by the same company are commonly started only if they show a higher expected return, even though there would be little difference in the cost of capital. The higher expected return is a manifestation of the desire for no more than 10% of the "tail" of the probability curve to fall to the left of the cost-of-capital line.

Table 14.2 sets out the sequential steps in decision making for investment in major resources projects.

VALUE AS A FUNCTION OF RISK

The relationship between risk and return for *financial* instruments has been well understood since at least the mid-1960s, when economists developed the capital asset pricing model. (Original work on portfolio diversification to reduce the standard deviation of returns is credited mainly to Markowitz [1952]. Initial work on the capital asset pricing model is credited to Sharpe, Lintner, and Treynor; see, e.g., Sharpe [1964] or Lintner [1965].) This model states that, in a competitive market, the expected risk premium varies in direct proportion to the variability of the instrument compared to the market. This model has proven to be most successful for stocks and bonds, and particularly for financial instruments that have well established records of variability and for which a *competitive* market exists.

Unlike financial instruments, individual capital investments by firms cannot rely on any well-established record of variability since each investment is unique. Some elements of variability can be understood (e.g., mineral commodity prices) and certain *types* of investments may correlate with previous investments of the same type, but unlike with stocks and bonds, past performance is only a weak indicator of future variability of individual new investments. In

TABLE 14.3 Base data for discounted average cost example

Base Data	New Dragline	Old Dragline
Estimated production rate, m^3/*operating hour*	2,000	2,250
Initial capital cost (written off over 10 years), $	30,000,000	12,200,000
Refurbishment cost (written off over 10 years), $	—	10,300,000
Total initial capital cost, $	30,000,000	22,500,000
Estimated production time available for dragline in a year and standard deviation of hours, *operating hours/year*	6,600±400	5,867±650
Operating cost (including labor), *$/operating hour*	700	1,115
Type of depreciation	Straight line over 10 years	Straight line over 10 years
Corporate income tax rate	50%	50%
Machine or mine life	15 years	15 years

addition, unlike with stocks and bonds, at the time the decision is made the value of *new* investments does not have the benefit of a competitive market assessing its value over time.

This section illustrates how the uncertainty criterion can be used to assess the appropriate premium for individual elements contributing to the uncertainty of an individual investment. Results are not derived from the capital asset pricing model, but they are not inconsistent with results that might flow from such a model were it possible to apply to individual unique investments.

A Case Study: New or Old Equipment?

The example sets out a simplified case study of two alternative dragline purchases. The first case—a new machine—has high capital costs, low operating costs, and availability that is subject to little uncertainty. The second case—an old machine—has lower capital costs, higher operating costs, and availability that is much less predictable. It is the reliability of the machine (variability in availability) that is of interest in this case study. Both machines move the same amount of waste annually.

The firm has sufficient capital to pursue the higher-capital-cost option, and both cases will be assessed on the assumption of a required 15% DCF return on investment. The particulars are set out in Table 14.3. (In this chapter, all volumes are expressed as bank, or unswelled, quantities.) The discounted average cost is the price *per unit of production* you would have to pay someone else *with the same investment criteria as yourself* to have the production undertaken.

The results of the tabulation are set out in Tables 14.4 and 14.5.

Deterministic DCF Results When the original question was posed, the capital and operating cost alternatives were deliberately chosen so that the discounted average cost of waste removal would come to the same answer in both cases. In this example, a contract price of $0.93/m^3 would yield the investor a return of 15%. The cost is the same for both draglines.

TABLE 14.4 Discounted cash flow: new dragline

	Year							
	0	1	2	3		13	14	15
Annual quantity moved, *thousand* m^3		13,200	13,200	13,200	...	13,200	13,200	13,200
Revenue received at \$0.9323/m^3, *thousand \$*		12,306	12,306	12,306	...	12,306	12,306	12,306
Capital expenditure, *thousand \$*	30,000							
Tax depreciation, *thousand \$*		3,000	3,000	3,000	...		Nil after year 10	
Written-down value at end of year, *thousand \$*		27,000	24,000	21,000	...		Nil after year 10	
Annual operating cost, *thousand \$*		4,620	4,620	4,620	...	4,620	4,620	4,620
Profit before tax, *thousand \$*		4,686	4,686	4,686	...	7,686	7,686	7,686
Tax payable, *thousand \$*		2,343	2,343	2,343	...	3,843	3,843	3,843
Net cash flow, *thousand \$*	(30,000)	5,343	5,343	5,343	...	3,843	3,843	3,843
Discount factor	*1.0000*	*0.8696*	*0.7561*	*0.6575*		*0.1625*	*0.1413*	*0.1229*
Discounted cash flow, *thousand \$*	(30,000)	4,646	4,040	3,513	...	625	543	472
Net present value, *\$*	0*							

Note: All numbers in parentheses indicate negative values.

* Unit revenue (line 2 of table) was deliberately chosen to equate NPV to zero

TABLE 14.5 Discounted cash flow: refurbished dragline

	Year							
	0	1	2	3		13	14	15
Annual quantity moved, *thousand* m^3		13,200	13,200	13,200	...	13,200	13,200	13,200
Revenue received at \$0.9323/m^3, *thousand \$*		12,306	12,306	12,306	...	12,306	12,306	12,306
Capital expenditure, *thousand \$*	22,500							
Tax depreciation, *thousand \$*		2,250	2,250	2,250	...		Nil after year 10	
Written-down value at end of year, *thousand \$*		20,250	18,000	15,750	...		Nil after year 10	
Annual operating cost, *thousand \$*		6,542	6,542	6,542	...	6,542	6,542	6,542
Profit before tax, *thousand \$*		3,515	3,515	3,515	...	5,765	5,765	5,765
Tax payable, *thousand \$*		1,757	1,757	1,757	...	2,882	2,882	2,882
Net cash flow, *thousand \$*	(22,500)	4,007	4,007	4,007	...	2,882	2,882	2,882
Discount factor	*1.0000*	*0.8696*	*0.7561*	*0.6575*		*0.1625*	*0.1413*	*0.1229*
Discounted cash flow, *thousand \$*	(22,500)	3,485	3,030	2,635	...	468	407	354
Net present value, *\$*	0*							

Note: All numbers in parentheses indicate negative values.

* Unit revenue (line 2 of table) was deliberately chosen to equate NPV to zero

Figure 14.4 shows the cost structure of the two cases, plotting operating cost and equivalent capital cost. The equivalent capital cost includes repayment of capital, taxes, and accounting "profits."

Conventional DCF analysis does not allow these two cases to be differentiated. The new dragline costs an additional \$7.5 million, but this extra expenditure also yields 15% return. Unless the marginal cost of this extra \$7.5 million of capital were higher than the cost of capital for the first \$22.5 million of expenditure, the NPV of the two cases would also show little difference.

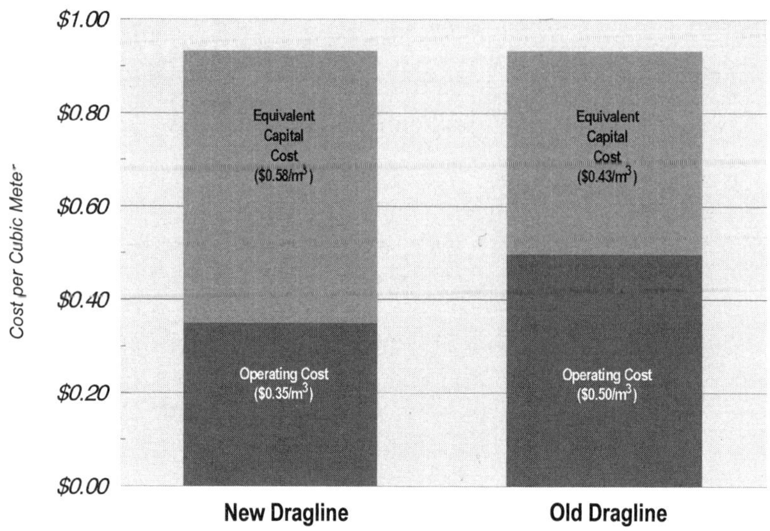

FIGURE 14.4 Discounted average cost: capital and operating cost mix

By focusing on *expected* values, this analysis overlooks the fact that availability is more variable in the older machine than in the new machine. This is the prime reason that older machines are less favored—they are less dependable. Even if *on average* the output is the same, people are prepared to pay a premium to avoid the risk associated with the older machine. (Whether they *should* be seeking a premium in this case or not will be discussed later.)

This sort of analysis can be made only through a simulation technique.

DCF Simulation A Monte Carlo simulation was undertaken, varying the actual number of hours of available time for each dragline for each year. With varying hours each year, adjustments were made to hourly maintenance and operating costs, including an adjustment for changes in electricity demand charges.

Simulations were undertaken in two ways.

The first way assumed that a contract is available for earthmoving by either of these machines at a contract price of $0.93/m^3$. Under this scenario, variations in availability translate into variations in the rate of return.

The results are set out in Table 14.6 and plotted in Figure 14.5. Minor changes occur to the *expected* return on investment, but fairly substantial differences are evident in the uncertainty criterion. If the two alternatives represented complete projects and the distribution represented *all* the uncertainty, then the shaded areas could be reconciled against the marginal cost of capital and a choice made. For example, at a cost of capital of 10%, both alternatives satisfy the uncertainty criterion, and the choice should be made based on NPV.

TABLE 14.6 Probabilistic discounted cash flow results

	Expected Result	Standard Deviation	Minimum Result	Maximum Result
New Dragline				
Discounted average cost, $	0.9358	0.0507	0.8115	1.1867
Annual operating hours	6,597	410	4,992	7,791
Net present value (at 15% DCF rate of return), $	(13,236)	1,896,600	(7,426,200)	5,502,900
Hourly average operating cost, $	701.83	26.69	636.47	833.81
Annual production, m^3	13,194,300	821,100	9,985,000	15,582,300
Return on investment, %	*14.97*	*1.39*	*9.23*	*18.87*
Refurbished Dragline				
Discounted average cost, $	0.9439	0.0969	0.7503	1.6390
Annual operating hours	5,859	674	2,990	7,798
Net present value (at 15% DCF rate of return), $	(34,902)	3,258,500	(13,900,800)	9,334,800
Hourly average operating cost, $	1,125.65	88.45	948.99	1,759.79
Annual production, m^3	13,183,750	1,517,200	6,727,500	17,546,500
Return on investment, %	*14.85*	*3.28*	*(2.63)*	*23.45*

Note: Numbers in parentheses indicate negative currency values.

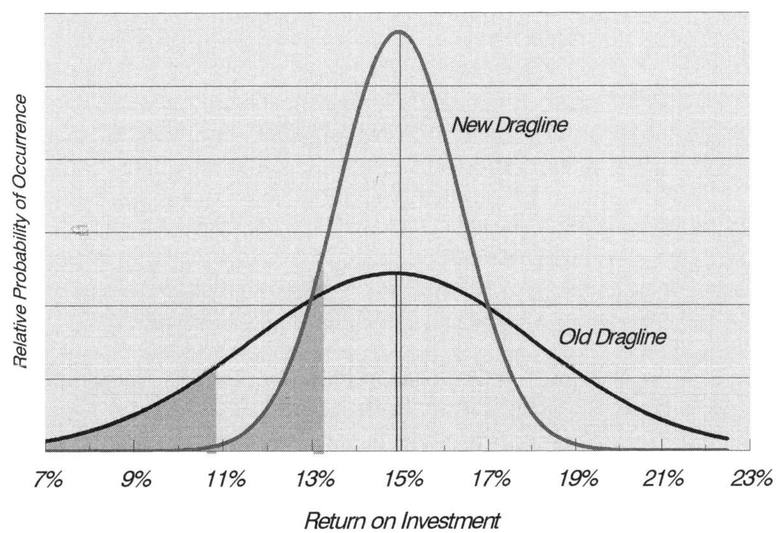

FIGURE 14.5 Effect of variability on return on investment

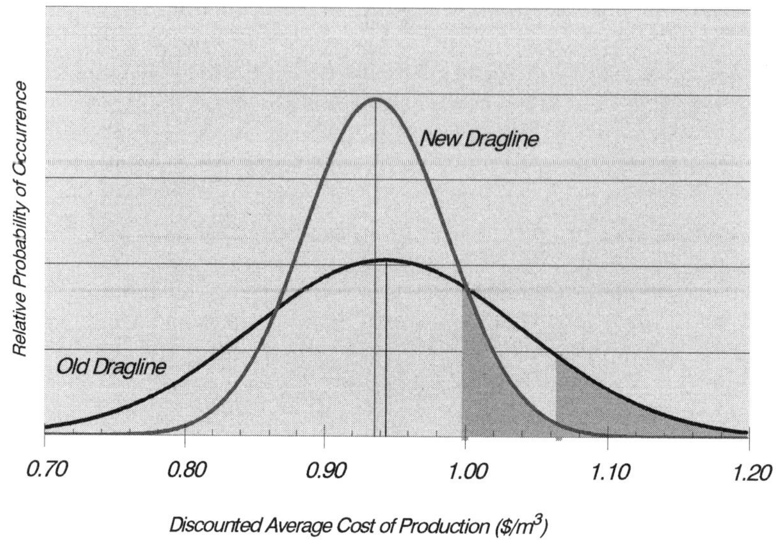

FIGURE 14.6 Effect of variability on cost of production

Unfortunately, seldom are *all* elements contributing to uncertainty known. A more useful result is available if the change in price can be related to just one such element on its own. This is the objective of the second type of analysis.

Decisions Based on the Uncertainty Model In this second type of analysis, the question is asked in a reverse fashion: What would the selling price (discounted average cost of production) have to be for 90% confidence that the investment will yield a return exceeding the required return? Variations in availability under this model yield variations in the contract price necessary to achieve the 15% return.

Figure 14.6 shows the results of the simulation, plotting the discounted average cost of production needed to achieve the desired 15% return on investment. The 90% confidence level (leaving 10% of the area in the right-side tail of the curve) is equivalent to 1.29 standard deviations.

In Figure 14.6, 1.29 standard deviations is represented by

New dragline (mean $0.9358, standard deviation $0.0507)	$1.00/m^3
Old dragline (mean $0.9439, standard deviation $0.0969)	$1.07/m^3

The old dragline requires a premium of approximately 7%, or $0.07/m^3, before it can be considered an equivalent investment from this uncertainty perspective.

This is a definitive answer and depends only on the estimate of variability in availability. It allows quantitative values to be placed and objective comparisons to be made on intuitively understood uncertainties.

Should this 7% "risk" premium be used, or should the NPV value be used? In this case, the new dragline had a higher value when the uncertainty criterion was used and an equal value when the NPV criterion was used, so there is no advantage in purchasing the old dragline. If the extra capital requirements of the new dragline meant a higher cost of capital, then the two criteria would have pointed to different recommendations.

The uncertainty criterion is a precondition. It applies only until satisfaction of an organization's tolerance for risk. Diversified enterprises can sustain a higher tolerance for risk on individual investments than less diversified enterprises. Thus, if the (old) dragline is the *only* dragline owned by the company and it represents the only source of revenue, then the uncertainty criterion must be rigidly applied. The old dragline should be purchased only if it has a discounted average cost of production (in expected NPV terms) 7% higher than for the new dragline.

If there are multiple machines on-site and the performance of one machine over a short period is not crucial, then the uncertainty precondition can be assumed satisfied. The selection can be made by using only the NPV criterion.

THE VALUE (AND DIFFICULTY) OF PROBABILISTIC ASSESSMENT

To identify the variability of just one input is usually a fairly straightforward task. Personnel experienced in one particular area can usually develop consensus as to the variability of certain components.

For whole-project analysis, requiring multiple interacting variabilities, the problem expands substantially. The difficulty in estimating the underlying probability in these sorts of assessments is acknowledged, but this does not mean that meaningful knowledge cannot be gained even with this limitation. Probabilistic methods fill two important functions that cannot be addressed easily in any other way:

1. They provide a mechanism for personnel who *understand* any element of uncertainty to quantify this element. Individual subjective or objective assessments can be separately defined but collectively analyzed. The discipline imposed on individual skilled team members to consider uncertainties in their area of knowledge frequently results in substantial changes and improvements in the robustness of plans. This knowledge often cannot be drawn out and assimilated in any other way.
2. There are certain elements that are incorrectly portrayed in *any* deterministic analysis. Using a deterministic variable is tantamount to assuming *no* variability—a case that may result in a systematic error. Even an *assumed* underlying stochastic characterization will commonly yield more reliable results than a deterministic assessment that assumes no such variability.

The following case study, adapted from Runge (1994), illustrates one use of the technique in valuing a project for equity participation.

A Case Study: Equity Valuation

Valuation of projects for multiparty equity participation is a particularly appropriate application of stochastic financial evaluation techniques. Every project is subject to varying forms of risk, and one of the prime objectives in multiparty ownership derives from the differing contribution each party makes to offsetting these risks.

This project assumes a proposed joint-venture ownership structure common in Australian mining. The major contributor and project promoter usually becomes the operator in the venture. Apart from the monetary contribution, this party also has the skills and knowledge for objective decision making concerning *production.* Operators have the capacity to provide guarantees of production that other participants cannot offer from their own resources. A financier is usually the second party in a joint venture. Financiers provide access to capital and feedback to debt markets in a way that other participants cannot offer from their own resources.* The third party (in many joint ventures or equity arrangements) contributes by way of understanding market risks and typically has the capacity to provide some guarantees regarding market offtake.

Anyone who buys a large amount of some product annually from a variety of sources can provide a guarantee to purchase, say, one-tenth of this annual requirement from one particular source at essentially no cost, assuming the price is market determined. Few customers will provide such a guarantee to a supplier without some value in return. Equity participation has frequently been one area for suppliers to offer value in return.

A guarantee of offtake by a customer who will be in the market to buy the product anyway essentially costs the customer nothing. On the other hand, the *loss of revenue* incurred in the absence of a sale may have a large impact on project economics and represents a risk. Indeed, in a variable market, guaranteed or underwritten sales at *below market* price are valuable options to hold. Accountants in large companies are increasingly wary of the contingent liabilities (and doubtful value) associated with options. Nevertheless, it is the differential value of the guarantees that forms the basis of risk sharing and ultimately the valuation of individual equity stakes over the notional net present value of future cash flows. This is the basis of the analysis described in the following paragraphs.

This original study was undertaken for a coal producer in Australia and was simulated on a financial model set up on a spreadsheet. The model has been simplified for this example, but it nevertheless captures the essence of the analysis, which is applicable to a wide range of similar problems.

* The involvement of a financial institution as an equity participant is a strong signal to other financial institutions about risks. These other institutions assume that this finance equity participant undertakes appropriate due diligence before funds are committed, and they free ride on this information to minimize their own work when assessing their own involvement. Thus, the market confidence provided by such a participant cannot be replicated by anyone except an institution seen to be diligent and seen to have objectives similar to institutions providing debt funding.

This particular project came into the owners' hands only after a large amount of money had been spent on assessing many prior unsuccessful opportunities, plus about \$10 million on acquiring the project itself. The owners were considering selling 10% equity in the project to a potential large customer in return for locking in sales and removing the uncertainty concerning mine offtake. Two questions had to be addressed:

1. The first question concerned pricing of a 10% equity stake. Should a 10% equity stake for a new participant be priced at \$1 million, or should it be priced at some larger amount to reflect the cost of the unsuccessful exploration efforts over the previous years? With no other contribution except money, the market price for a 10% equity stake in such a near-proven project would be much higher than \$1 million. On the other hand, a large customer *could* make a contribution beyond the monetary one—by guaranteeing offtake.
2. The second question regarded these offtake guarantees. Existing partners had the resources to guarantee only about 60% of the offtake—either by delaying start-up until contracts for this offtake were in place or by transference of existing contracts. Access to additional markets by way of guaranteed or underwriting sales contracts is certainly valuable. But how can these guarantees be valued?

Assuming all the output could be placed at current prices (100% sales at contract prices), the project appeared quite viable.

Conventional (deterministic) cash flow analyses assume that whatever coal is produced will be sold. This is a correct assumption. Most mines are capital-intensive. Once capital has been sunk, the marginal return from selling coal cheaply is still much better than not selling it at all. Placing a value on an offtake guarantee requires the base case to have less than a 100% chance of selling the offtake or, at least, some substantial discount on sales that cannot be made into long-term markets.

For the base case probabilistic analysis, the assumption was made that all the output could be sold. Quantities not precommitted were assumed to be subject to the success of contract negotiations undertaken annually. Contract negotiations were modeled with uncertainty. Substantially lower prices were assumed for sales on the spot market when contracts were unavailable. The uncertainties used in the simulation are defined in the next three subsections. The base case probabilistic definition of market offtake is set out in the first of these subsections.

Probability of New, Long-Term Contracts A base case analysis has to be prepared with uncertainties sourced from marketing personnel. These uncertainties do not fit any neat probability distribution, but, fortunately, simulation methods do not necessarily require such distributions.

For this case it was assumed in the model that three serious opportunities for new coal supply contracts would present themselves annually. Since coal sold under contract is priced with some consistency (i.e., priced according to industry benchmarks), the probability of obtaining *contract* sales is only weakly related to a willingness to discount price. Initial-year discounts are possible as a precursor to fully priced contracted sales in the following year.

For this model it was assumed that for any one opportunity presented in each year, there was a 10% chance of success. A maximum of one new contract per year was also assumed. Contracts were assumed to run indefinitely, once awarded, although it would have been a simple matter to model expiration (and rewinning) of existing contracts as well based on any assumed criteria within the experience of the marketing personnel. Contracts were assumed to vary in size anywhere from 100,000 t annually to 500,000 t.

It was assumed that production not sold on a contract basis would be placed on the spot market at a $10.00/t discount–a sale price that exceeded the marginal costs of production but that rendered those tons unprofitable when fixed costs were apportioned against them (the conventional accounting way of reporting profit).

Offtake by New Equity Participant If a new participant is to provide offtake guarantees rather than pay full market price for its equity position, then this guarantee must have value. The purpose of the assessment was to determine this value.

Guarantees for annual outputs ranging from 0 to 50% of mine output were considered. (The existing owners would not have sold any equity to a customer without a guarantee amounting to at least 25% of mine output. A customer with full access to mine financial data can use this information to gain lower prices, and the gain from this lower price can exceed the small loss incurred on its equity investment.)

Figure 14.7 shows the results of the analysis comparing the increase in net present value of the project (over the base case with no such guarantees) with increasing guaranteed offtake. Net present values shown are *mean* values determined after 100 or more stochastic recalculations of the cash flow for each point on the graph.

At a 15% discount rate, the project base case (with uncertainties in market offtake) showed a net present value of approximately $32 million after including all exploration, land acquisition, and capital expenditures. In other words, a mining company seeking a (real) return of 15% could afford to pay all past and future direct costs and up to $32 million just for the ownership rights to the project.

From Figure 14.7, if the ownership rights *include* a guarantee for 25% of the project offtake at market prices, then this adds approximately $8 million of value. Unless the new participant chooses to pay a direct premium, this kind of guarantee implies a maximum equity stake for the new participant of 20% ([$32 million + $8 million] × 20% = $8 million)–a level at which the value that the new participant adds to the project through the offtake guarantees equates to the participant's proportion of the (new) total project value.

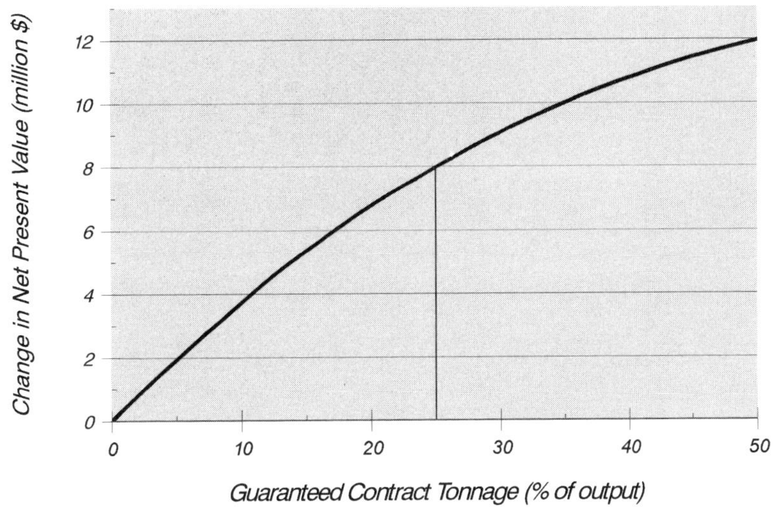

FIGURE 14.7 Change in NPV with change in contracted offtake

To achieve this result, the new equity participant does not have to directly buy the mine output; it must only guarantee the sale. Nevertheless, the two are not necessarily equivalent. Without an actual purchase, the tonnage may still be competing in the marketplace—perhaps in competition with the mine's own sales personnel seeking to place the balance of the uncommitted tonnage. In this case, the changed probabilities of residual offtake would have to be incorporated in the simulation.

In making the preceding assessments, the owners were quite aware that the one factor affecting mine profitability the most in the early 1980s was shortfalls in delivered quantities early in the mine life. The time it takes to find and consolidate long-term supply relationships with customers is frequently underestimated, and until this consolidation has been achieved there is the continuing risk of offtake shortfalls. Though not common in the coal industry, this sort of situation is frequently handled in other industries through underwriting arrangements. One further analysis was undertaken to examine the value of such an arrangement.

Underwriting Option The underwriting option was an alternative considered on top of the long-term supply contract. The particular value of this option related to minimizing the risk over the critical first 5 years of full production for the project. At the end of the 5-year period, it was considered that there would be a sufficiently high probability that long-term contracts would be in place that such an agreement would no longer be necessary.

Underwritten quantities from 100,000 tpy to 600,000 tpy were examined and assumed a discount on market price of $2.50/t. Since this tonnage was just residual output pending commitment to long-term contracts, its value to the project was primarily one of cash flow rather than return. The change in net present value with change in underwriting tonnage is shown in Figure 14.8.

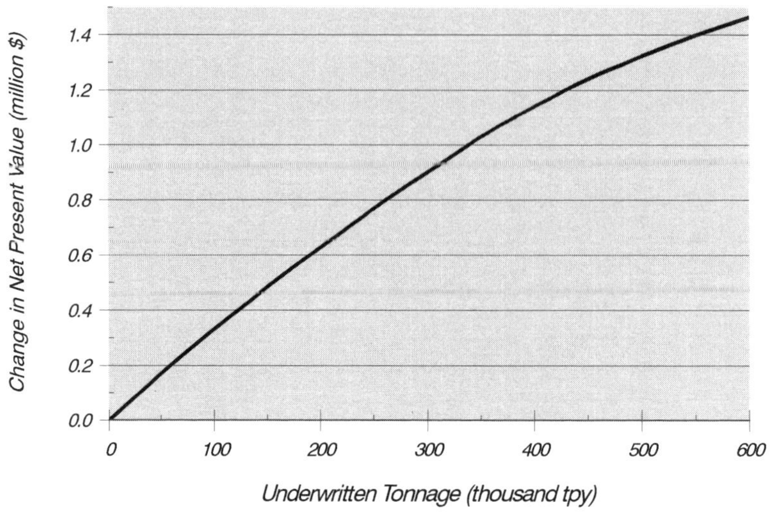

FIGURE 14.8 Change in NPV with size of underwritten tonnage

Compared to the long-term offtake guarantee at market prices, a limited-term underwriting agreement at discounted prices is naturally of less value. This is reflected in the change in *expected* NPV shown in Figure 14.8. Nevertheless, this modeling also highlighted possible short-term cash flow restrictions and the adverse effects on decision making in cases of offtake shortfalls early in the mine life—aspects too complex to allow easy summarization in this example.

Without an underwriting agreement, the existing owners effectively take the risk on placement of this output themselves. Similarly, from the underwriter's point of view, the potential that no tonnage will be taken up must be balanced against the risk that all the tonnage will be taken up. Table 14.7 sets out the probabilities and average quantities calculated after 500 simulations of the cash flow, assuming a 5-year underwriting agreement for 500,000 tpy.

PROBABILISTIC ASSESSMENTS WITH UNCERTAIN INPUTS

The ultimate test of any evaluation technique is the value of the results that it provides. Although the preceding example and other studies undertaken by the author have frequently had to use stochastic inputs for which the characteristics are quite poorly understood, in many cases the results turn out to be quite robust *despite* the imprecision of the inputs. There is a reason for this. The value of the modeling is in its treatment of the *interrelationships* among variables. The modeling of interrelationships primarily requires variables to change; it is less important whether their variability is characterized by a normal distribution, lognormal distribution, or any other type of distribution.

This conclusion does not apply universally. The preceding case study looks at the *change* in return on investment with *change* in one of the model inputs—in this case, a model input that can be thought of only in some stochastic way. A

TABLE 14.7 Quantities and probabilities of market uptake

Probability or Quantity Description	Value
Probability of zero tonnage being taken up	9%
Average tonnage taken up (total quantity over 5-year life of agreement)	1.2 million t
Probability that maximum tonnage will be taken up	12%

whole-project analysis, for example, would need all stochastic variables faithfully modeled to allow a probabilistic assessment of the return on investment such as shown in Figure 14.2. This is not a trivial task. Since very few decision makers know how to interpret the results, the value is currently unlikely to be enough to pay back the effort.

This does not mean that whole-project assessment should not be undertaken. The technique models all the important variables simultaneously, and it is possible to use the model itself to determine whether the results are sensitive to the *characteristics* of the input. Even if the characteristics of an input are unknown, the model can be simulated over a *range* of inputs. If the results of the model change, then this is a signal that better definition of the input is necessary. The model itself is an invaluable guide to understanding which parts of the underlying plan (with uncertainty) translate most into uncertainty in the result. These are the parts that clearly need to be understood the best.

This highlights the second and perhaps primary value from probabilistic analysis—the value from understanding the problem better. In the preceding example, many practitioners could argue the logic used for modeling contracted market offtake. Nevertheless, before the probabilistic assessment, the previous deterministic assessment assumed 100% success in placing all the mine output from the first day. Compared to the primitiveness and unwitting optimism of the deterministic method, the probabilistic logic looks thoroughly professional. Only in the modeling process is there a mechanism to capture and document the logic that drives many of the participants in a study. When this logic is brought out into the open, there is potential for contribution, improvement, and scrutiny by others on the project team, resulting in easier auditing and improved decision making—both of which result in risk reduction.

CHAPTER 15

At-Risk Discounted Cash Flow Analysis

What does it mean to "invest" in something?

The dictionary defines *investment* as an act of putting money to use by purchase of or expenditure on something offering profitable returns. Popular usage is not quite so broad. Ordinarily, funds placed in a savings account are not considered an investment even though they do offer profitable returns. Conversely, the same funds placed in the stock market in a company that has never paid a dividend *are* regarded as an investment.

The essential difference in popular usage of the term has to do with risk. Money is *placed* in a savings account, not *invested* in a savings account, because none of it is at risk. There is no uncertainty about getting the original sum back. Some "investors" talk about investing in mutual funds when their funds are available on demand. They may even have a checkbook to draw from. This sort of funds placement has a different risk-and-return profile, but for practical purposes the risk is quantifiable and even insurable. It is just one point on a well-defined risk-return line.

The essential characteristic of an investment is that some or all of it is at risk or is unavailable until some time in the future. As Lachmann (1978) says: "Money is an asset, but it is not a capital good like other elements of a production plan" (p. 87); and "after all, one cannot earn a profit on capital without 'investing' it, and that means to dehomogenize money capital" (p. 36).

The decision to invest is a decision concerning the potential *loss of value* associated with changes that might take place between the time when the entrepreneur "invests" (or dehomogenizes) his or her money capital until the time when he or she is in a position to recover the capital in a "free" form (i.e., as money).

The at-risk capital approach explicitly aims at understanding the component of capital in an investment that is dehomogenized. It looks at two things: the loss of value and the time. This chapter sets out explicit tools to understand these two investment characteristics.

ROLE OF COST IN A DECISION

In Chapter 4, a distinction was made between (1) the cost of a whole project or event and (2) the cost associated with the decision. Capital decisions are recognized as long-term decisions, but once this path into the future is set upon it is not necessarily irrevocable. It is only the irrevocable part that constitutes value or likely loss of value in the event of unanticipated obstacles to plan fulfillment. The example in this section highlights this distinction. It also illustrates the relationship between "value" as privately understood (in the mind of the decision maker) and "value" in a market-based sense. It illustrates why it is rational and *essential* that these valuations are *not* in one-to-one correspondence at various stages in the capital investment decision process.

Note: Very few models in the economics literature allow for differences between marginal values in the mind of individual decision makers and the value in a market-based sense. The reason for this is simple: If such differences persisted for long, the individuals would buy or sell the goods in question and make a profit, and they would keep doing so until there was a close correspondence between the two valuations. In the consumer market or in the trading of stocks and bonds for example, where transaction cost is low and information is readily available, this is an accurate characterization of exactly what happens. For capital investments of a one-off nature, however, or for mineral commodities that are early in the chain of production, the relationship is not so clear-cut. The "market" can assimilate only knowledge that is generally available (known production capacities and cost, known demand, etc.), and participants in this market also realize that there are many factors that influence supply and demand that they are not aware of. Thus, market values may be inconsistent with underlying supply-and-demand economics for considerable amounts of time and may readjust dramatically once participants become aware of and recognize the impact of hitherto overlooked or unavailable information. Pindyck (1993) studied the present value model of commodity pricing by examining the joint dynamics of spot and futures prices and found "close conformance to the model for heating oil, but not for copper or lumber, and especially not for gold" (p. 529). This is consistent with minerals industry experience. The difference relates to the free availability of information, the costs of storage, and the ease of bringing additional supply onto the market. The model in the balance of this chapter explicitly separates private valuation—where a decision maker has, or assumes he or she has, superior information relative to the market in general—and market valuations.

Assume you are considering the purchase of a new car. Since this is a long-term investment, you are examining this proposal by using all the principles set out in this book. The purchase of the car has one great advantage over a mine investment—at any time the market value can be readily established, and this is a valuable aid to decision making.

TABLE 15.1 Quarterly value from vehicle usage

	Quarter				Other Quarters	End of 5 years
	1	2	3	4		
Value received, $	2,270.83	2,062.50	1,854.17	1,645.84	1,541.67	1,541.67
Resale, $						20,000
Discount factor at 7.5% annual rate	*0.9821*	*0.9645*	*0.9472*	*0.9302*	*	*0.6966*
Present value, $	2,230.15	1,989.25	1,756.28	1,531.01	*	15,005.03
Net present value, $	41,183.29					

* Discount factors and present values are calculated on a quarter-by-quarter basis by using Equation 5.2. The discount rate of 7.5% (on an annual basis) corresponds to a rate of 1.824% on a quarterly basis.

After due consideration you determine the value you expect to get from the car on a quarter-by-quarter basis for the next 5 years. You expect more value in the first year because the car is new: There is a novelty in driving it, and you will visit places that you have been putting off visiting. After the first year, the novelty value will diminish and the value you will continue to derive will come directly from functionality. Table 15.1 sets out the quarter-by-quarter expected value, including the resale value at the end of 5 years. The amounts shown are net, after fuel, maintenance, and other operating expenses. (There is no suggestion that this procedure is actually suited for an investment decision concerning a private motor vehicle, the value of which, for most people, derives from subjective criteria such as driving pleasure, status, and prestige. The example simply aims to illustrate the steps in the procedure for any type of investment.)

The funds put into the car (or the monthly payments on the car) will inhibit other purchases that are worth something to you. You estimate the opportunity cost of this capital is 7.5%. Applying a 7.5% discount rate to the amounts in Table 15.1, including the trade-in value, results in a net present value of the car to you of $41,183. The purchase price of the car is $40,000, and since your net present value exceeds this amount, it is viable to proceed.

The quarter-by-quarter value received and the cumulative value received are shown in Figure 15.1, with the payback profile in Figure 15.2. In nondiscounted terms, there is a 5-year payback period—you do not get your money back until you finally sell the car.

If the purchase of the car locked you into a commitment to hold it for the full 5 years, then further analysis is probably not warranted. Yet purchases of this type are not irrevocable. You have an option *at any time* to sell the car. You can change the payback profile at any time. If you sold early, could you get your money back sooner? Figure 15.3 shows the estimated resale value of the car over the 5-year period.

Anyone who has had to sell a fairly new car recognizes the large loss in market value initially. A similar problem exists with all large capital investments. Buying and selling a house, for example, typically results in 5% immediate loss due to transactions cost—agency and legal fees, advertising, and the like.

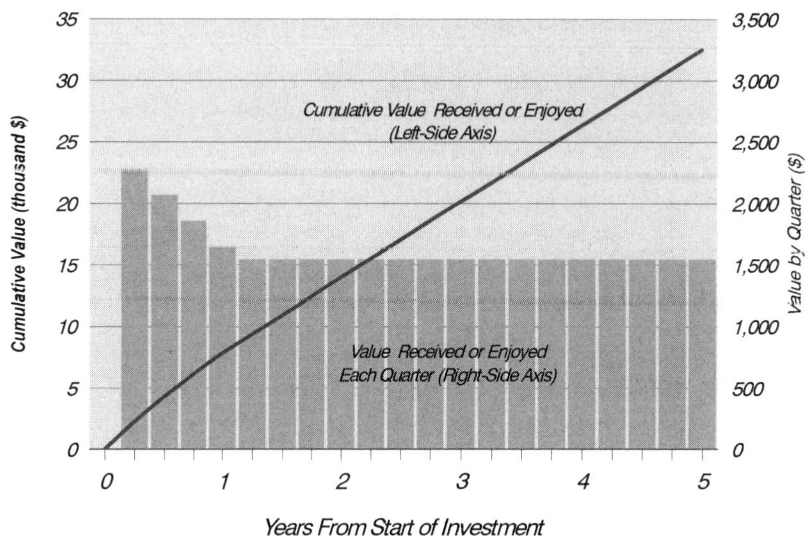

FIGURE 15.1 Quarterly and cumulative value from vehicle usage

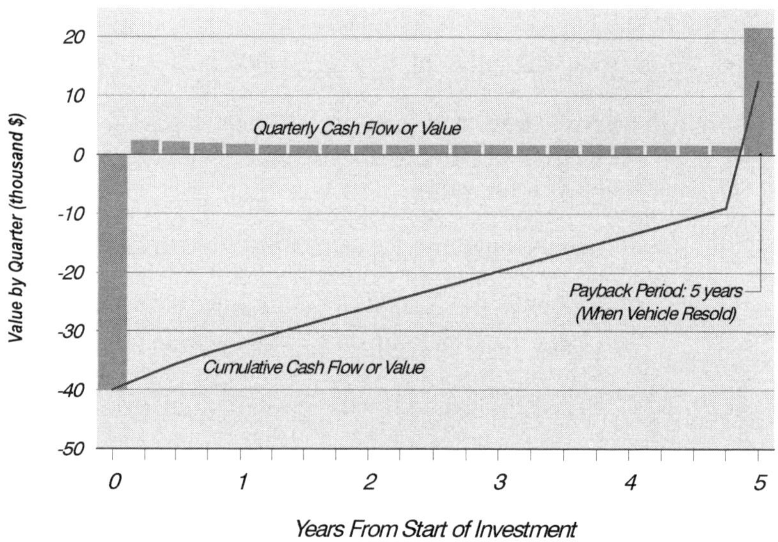

FIGURE 15.2 Vehicle purchase payback period, conventional approach

In this example, the day you drive the car out of the showroom, its market value drops by $5,000. Yet car sales are not subject to high transaction costs. Why should there be so large a difference between the market value of a new car and the market value of a 1-day-old car?

Many studies have examined this well-understood phenomenon (see, for example, Akerlof [1984]). The problem with motor vehicles and with most large capital investments is one of information asymmetry. If you valued the car at $40,000 (or more) on the day you drove it out and no damage has been

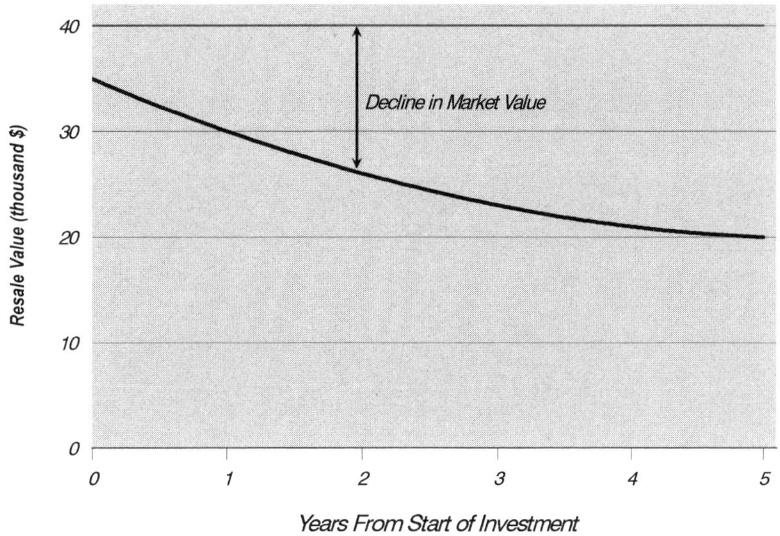

FIGURE 15.3 Vehicle resale value over time

done to it, then *your* (private) value will not change. But "the market" does not know that no damage has been done to it. The engine might have had no oil! There is no way for an outside observer to know—except by asking you—and you have an incentive *not* to tell the truth in the event that something did happen. This is an instance where the market price—for *your* car, at least—is actually *wrong* because it is ill informed.

Market-related trades occur daily, and when we make choices and engage in market interaction any difference between our valuation and the market valuation is *our* risk. The *real* risk is not the total commitment on the vehicle purchase but rather the difference between our value and the market value. If something unexpected happens and we have to sell, this is the loss we will incur. Initially the potential difference is quite large, but if your choice is a good one the value that you receive over time increases at a faster rate than the market price of your investment decreases. The difference is relevant when, or if, you exercise the option to sell.

Figure 15.4 shows the cumulative value gained from use of the car, as well as the loss of value if at any time the option to sell had to be exercised. The figure puts an entirely different perspective on the investment decision. The true commitment at any time is the difference between the value lost in the event of sale and the cumulative value gained. After 2 years, the owner is in a position to sell the car, and the value received over the 2-year period is equal to the loss in value on the sale.

Thus, after 2 years the capital appears to be still "tied up" (dehomogenized, in Lachmannian terms) in the car, but you are in a position to recover your capital in a "free" form (i.e., as money). The amount of money that is "at risk" is

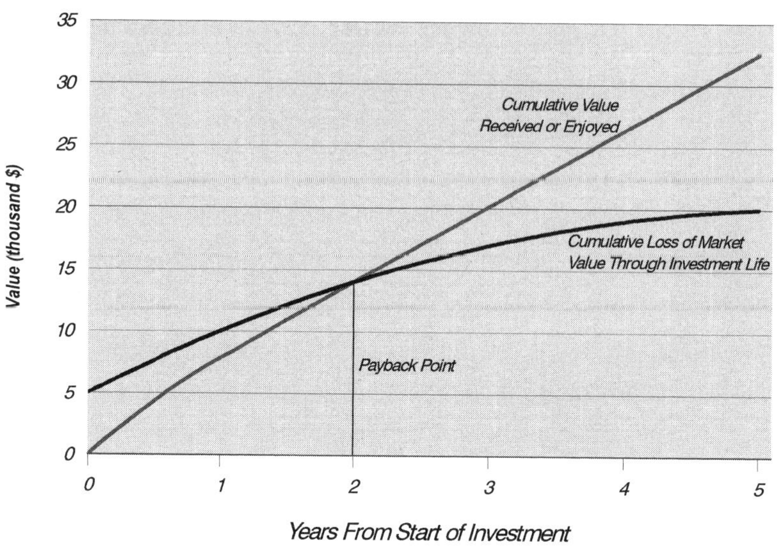

FIGURE 15.4 Payback from a loss-of-value perspective

the maximum difference between the two lines leading up to the payback period. After 2 years money is no longer at risk.

This concept of payback is directly related to the risk preferences of an actor. Further, as demonstrated in this example, there is not necessarily a strong correlation between discount rate and payback period. Most analysts would suggest that decision makers demanding payback in 2 years are quite risk averse. At the same time, decision makers who use discount rates of 7.5% are commonly considered not at all risk averse.

A crucial element in this view of risk is how well informed the market is, because this is what determines the shape of the market value line during the life of the investment. Decision making for new capital investments by firms is subject to the same market scrutiny, but it also *benefits* from this scrutiny. The benefit occurs in two areas:

- For projects proceeding according to plan, the better informed the market, the closer the market valuation will be to the internal-to-the-firm value. This reduces the capital at risk, as well as the time until payback.
- Markets convey important knowledge that may not be available to a company's own personnel. If the market judgment of the value of a project is different than the internal-to-the-firm value, then this may be due to knowledge that participants in the market at large have that the firm does not. Companies ignore these market signals at their peril, but the signals have less reliability if the market is ill informed.

On the one hand, underpinning an actor's valuation is his or her knowledge of the project and confidence in plan implementation. On the other hand, underpinning market valuation is what the "market" knows about the project–including knowledge not available to the actor–and its confidence in the actor's actions. Any reduction in the difference between these two values reduces the amount of the actor's capital that is at risk and minimizes the *time* that the capital is at risk.

CAPITAL AND REAL OPTIONS

The objective of investing in projects is to obtain a return. Indeed, *capital* is defined as the present *value* of these expected returns. In practice, this capital is deployed in many areas, each making differing contributions to reducing the risk, decreasing the operating costs, or providing some other efficiency. Since all of the capital comes from the same sources and must yield an overall satisfactory return, it tends to be treated alike.

Understanding the impact of this and ensuring that the resources allocated to capital are expended efficiently first requires examination of where the capital goes. "Capital allocation" here does not mean allocation in the physical sense (buildings, vehicles, and the like); rather, it means allocation in terms of the contribution in a logical sense.

This issue was introduced in Chapter 14 concerning the use of different discount rates for depreciation allowances. Summers (1987b, p. 31), following a survey of 200 large corporations, concluded that "[i]t is clear that the practice of separately discounting safe and unsafe components of a project's return, as suggested by theory, is a rarity in American industry." This section aims to explicitly identify a logical separation of capital on the basis of risk and to demonstrate the application of different discount rates to these logical subdivisions of capital in a mine.

Example 15.1:

Consider an expenditure of $50 million on a shaft to access an unknown orebody. This investment is worth nothing if the orebody proves to be grossly under expectation. Conversely, $50 million spent on a truck and loader fleet in the same circumstance may still be worth $40 million on the second-hand market. The amount of funds at risk is vastly different even though the total capital is the same.

The differences in the preceding example do not impact a traditional cash flow analysis since traditional cash flow analyses assume the project will proceed in accordance with expectations. The difference comes only for circumstances that are different from the assumptions used in the DCF—circumstances that require the option to abandon or some other option to be exercised. If the possibility of such circumstances were known prior to the DCF being undertaken, then options to allow for change can be built into the capital or operating structure and overall risk reduced.

The option to abandon is just one such option. However, an exit strategy of reselling capital at market prices is not the only fallback position and is seldom the most attractive alternative course of action. One reason such abandonment options are problematic is that at the time when this alternative is drawn upon, it is likely that other producers will be considering the same thing. The option is shared. Historical market values are not necessarily a reliable guide to market values forming part of an exit strategy.

Trigeorgis (1996) identifies six types of real options that decision makers can make use of, in addition to the option to abandon already discussed. (Real options are distinguished from financial options because they apply to capital assets rather than financial assets. For a more comprehensive analysis of the theory of real options, see, for example, Dixit and Pindyck [1994], Trigeorgis [1996], Pindyck [1988], Brennan and Schwartz [1985], and McDonald and Siegel [1986].) These seven types, summarized from Trigeorgis with comments applicable to mining investments, are as follows:

- Option to defer. This option compares the alternative of proceeding-now with the alternative of waiting-and-then-proceeding. These choices were noted as differences in *logical* projects in Chapter 12. With the arrival of information over time, the present value of deferred choices planned with greater efficiency with the (then) known information can exceed the present value of the same choice proceeded with immediately.
- Time-to-build option. This option involves a series of outlays and is very characteristic of mining investments. Each outlay sets in place production capability plus conditions for subsequent (reduced) outlays involving production. The payback example in Chapter 9 (see "Payback," p. 132) discussed one such time-to-build option.
- Option to abandon. This option has already been extensively discussed.
- Option to alter the operating scale. This option does not lead to more refined expectations, but it explicitly values alternatives that are less sensitive to imprecise expectations. A contractor who can be brought in and dismissed at will provides such an option. Well-established mining areas are home to many of these valuable options because in these areas independent contractors can diversify their sources of work. Under diversification there is little or no premium for contracting on or contracting off.
- Option to change (switch inputs or outputs). This option is particularly relevant in multiproduct mines, where different orebodies have different proportions of metals, or where (for example) coal can be washed to different specifications. It is valuable where there is imprecision in the expectations of the types and cost of inputs and outputs. Imbedded options to change are essential ingredients in long-life machines used for manufacture of fashion goods, for instance.
- Growth options. Similar to time-to-build options, these options recognize that certain actions (research and development, for instance) can be narrowly focused on the project at hand or more broadly focused. The option explicitly values the externalities from a more broadly focused approach where these externalities can be retained within the firm. Mining companies who learn the subtleties of treating certain ore types can retain and capitalize on this knowledge, leveraging it to gain advantaged positions in the ownership of other orebodies with similar characteristics.
- Multiple interacting options. This category is a recognition that combinations of options—e.g., one that offers protection against losses and one that offers revenue enhancement—can have a combined value exceeding the

sum of the separate parts. Gold mines frequently use multiple interacting options for sale of gold at average prices exceeding the market average price during the same period of production.

Options increase the reliability of achieving planned returns in addition to (frequently) improving the return on investment. A demonstrated history of capitalizing on this approach can also lower the financing cost of capital, further enhancing the return to shareholders.

The balance of this chapter sets out a mechanism and discussion of how capital in a project can be logically subdivided and assessed from this perspective.

UNDERSTANDING RISK THROUGH WORST-CASE SCENARIOS

In Chapter 14, probabilistic analysis was used as a tool to assess the probability of an unsatisfactory outcome from a decision. This is an important tool for decision making, but the question could legitimately be asked: Is it the *probability* of the bad outcome or the *magnitude* of the outcome that is of concern? Worst-case scenarios examine the magnitude of the bad outcome.

In the car example earlier in this chapter, the largest amount of capital at risk occurred immediately after the car was driven out of the showroom. The same situation occurs with most new major mining projects. At the time of commitment, the "market" knows least about the project, and until the mine is operating there are few mechanisms to judge if performance is according to expectations. If project development starts to go wrong, there are incentives for personnel involved to disguise this fact, at least until critical benchmarks fail to be achieved and it becomes evident. As with the market value for cars, the market value of a mining company during this period is not necessarily a correct reflection of the fundamental worth of the enterprise.

On the assumption that the greatest amount of capital at risk occurs relatively early in the project life, this section reassesses the role of the worst-case scenario, using it for understanding the need for built-in options.

Chapter 14 demonstrated the use of probabilistic analysis to show how low-value potential outcomes should be treated from a decision-making perspective. However, even in these analyses there is limited scope for understanding how much capital is at risk and how the logical project might change given change in one or more important inputs. Probabilistic analysis succeeds only in rolling back the changes as far as the underlying technical buildup—it offers very limited scope for changes to the mine plan in the face of changes in any of the variables in the model.

An example illustrates the problem. Sensitivity analysis (including probabilistic sensitivity analysis) might examine changes in the price of fuel oil that occur contemporaneously with an increase in the cost of labor. In any conceivable circumstance where these changes happen, there will also be a change in the selling prices of the mine outputs (perhaps some products increase in price, whereas others decrease in price).

In reality, if all of this happened, the mine plan would also change. Except in rare cases, mines always have the option to expand or reduce production and to trade in equipment for more suitable equipment. Resources would be redirected into producing more of the products for which the selling prices had increased and relatively less of the products for which the prices had decreased. If fuel oil prices rose substantially, equipment that operated on electricity would be more heavily used and equipment more dependent on fuel oil would be changed out and replaced by the electrically powered equivalent machine. Of course, the market value of no-longer-needed equipment in these circumstances might also be low—particularly if it is a type of equipment unique to the mining industry.

In the long run, it is the ability or *inability* of the project to adapt to these sorts of changes that establishes the real risk. Any comprehensive risk assessment must consider the ability of the project to accommodate this type of technical change. One can ascertain the value of this ability to change by the difference between the two cases: What would the market value of the project be if there was scope for change? What would the market value be if there was no scope for change?

The ideal technique is one similar to the car example earlier in this chapter. The at-risk capital and payback period should be calculable by tracking the market value of a mine under a changed circumstance and comparing this with what the market value might be under some other scenario. The technique set out in the following section is the first step toward this objective. There is no easy technique for plotting market value continually over time. However, for any one set of circumstances, the difference between planned (expected) value and market value is calculable.

The technique in the following section is aimed at dissecting the capital that goes into a project, determining how much of it is really at risk, and making judgments based on the return on the *risk* capital.

The logic behind the analysis is identical to the logic employed during the final stages of setting up a financial structure of a project. When financiers analyze projects from a worst-case perspective, they are looking to ensure that their capital locked into the project is still secure. In the case set out in this chapter, the objective is to help *select* projects in the first instance. The objective is to highlight at an early stage the potential risk areas and come up with a design best able to accommodate the likely change. It also focuses attention more directly on the *risk* capital rather than the whole capital.

The at-risk capital approach does not aim to be as sophisticated as the probabilistic methods discussed in previous chapters, although there is no difficulty in combining the two approaches to address an expanded range of risk-based issues. The at-risk approach aims to objectively focus attention on the worst-case scenario to do the following:

1. Force critical examination of this scenario from a management viewpoint (if it happens, what should management do?) and ensure that operational, management, financial, or marketing arrangements do not inhibit

TABLE 15.2 Base data for at-risk capital calculation

Base Data	Case A	Case B
Estimated production rate, *units/year*	10,000	10,000
Initial capital cost (written off over project life), $	200,000	100,000
Operating cost per unit of production, $	1.915	5.957
Estimated selling price	$10.00/unit	
Depreciation over project life	Straight line over 4 years	
Corporate income tax rate	35%	
Long-term corporate financing cost	10% (equity + debt, after tax)	
Opportunity cost of capital (required return)	15%	

management's ability to react to this circumstance. Many of the real options listed and described earlier in this chapter are imbedded in the project even without deliberate design and are commonly not obvious. If overlooked they can be inadvertently foregone in the midst of negotiations for industrial, financial, environmental, government, and supply agreements.

2. Provide a mechanism for project selection based on the return on *risk* capital, not just return on the total capital. In this sense, the at-risk approach allows individual project cash flows to be conceptually committed in the same way that derivatives markets allow financial instruments to assign cash flows, i.e., by dissecting cash flows into risk and nonrisk tranches and valuing each element differently.

The following example study is limited to just financial considerations in the worst-case scenario; however, the normal use (and most valuable use) of the technique is when technical changes to the mine plan are also included.

EXAMPLE STUDY: AT-RISK CAPITAL CALCULATION

The example compares two alternatives from an at-risk capital perspective. The base data for the two cases are set out in Table 15.2. The operating and capital costs in the two cases have been explicitly chosen so that in this example the discounted average cost of production and return on investment are the same at the estimated base case selling price.

There are three steps in undertaking the analysis:

1. Determine the cost of production and return on investment for both cases assuming everything is according to plan.
2. Examine a worst-case scenario or any other scenario that reflects the opinion of outsiders who are judging the decision. Determine what the market value of the project is under this scenario. The difference is the amount of capital at risk from this perspective.
3. Reexamine the original scenario to see if options can be exercised or built in to minimize the loss of value. Dissect the original capital into at-risk and nonrisk tranches. Apply a nonrisk discount rate to the nonrisk tranche, and use the balance of the cash flows to determine the effective return on the capital that is at risk.

TABLE 15.3 Deterministic base case cash flow for case A: capital-intensive method

	Year				
	0	1	2	3	4
Production		10,000	10,000	10,000	10,000
Unit revenue, $		10	10	10	10
Total revenue, $		100,000	100,000	100,000	100,000
Capital cost, $	200,000				
Claimable depreciation, $		50,000	50,000	50,000	50,000
Unit operating costs, $		1.915	1.915	1.915	1.915
Total operating costs, $		19,149	19,149	19,149	19,149
Taxable profit, $		30,851	30,851	30,851	30,851
Tax payable at 35%, $		10,798	10,798	10,798	10,798
Cash flow, $	(200,000)	70,053	70,053	70,053	70,053
Discount factor at 15%	*1.000*	*0.870*	*0.756*	*0.658*	*0.572*
Present value, $	(200,000)	60,916	52,970	46,061	40,053
Net present value, $	0				

Note: Numbers in parentheses indicate negative values.

The three steps are set out in the text and tables of the next three subsections. The example is drawn from Runge (1994).

Conventional Analysis: Return on All of the Capital Invested

The first step in an at-risk capital dissection is to undertake a conventional deterministic discounted cash flow analysis. The simple discounted cash flow models are set out in Tables 15.3 and 15.4. The discounted cash flows are (barring the gross simplifications) identical to the kinds of cash flows undertaken daily by almost any investment analyst. The net present value of the project in both cases is zero. The initial capital has yet to be spent.

From a traditional payback perspective, both projects are identical. The payback profile for case A is shown in Figure 15.5. This case is more capital-intensive and has a higher maximum negative cash flow, but the payback period is the same as for case B (2.85 years).

Worst-Case Analysis: Determining the Nonrisk Capital

The second step in an at-risk capital dissection is to prepare a worst-case scenario. This scenario presupposes that initial capital has been committed. These scenarios can and should be prepared for any envisaged situation at any time through the project life. The objective is to pick a point where there is likelihood of the greatest divergence between (1) expected internal-to-the-firm value as shown in the base case cash flow and (2) value as assessed under the worst-case criteria.

TABLE 15.4 Deterministic base case cash flow for case B: less capital-intensive method

	Year				
	0	1	2	3	4
Production		10,000	10,000	10,000	10,000
Unit revenue, $		10	10	10	10
Total revenue, $		100,000	100,000	100,000	100,000
Capital cost, $	100,000				
Claimable depreciation, $		25,000	25,000	25,000	25,000
Unit operating costs, $		5.957	5.957	5.957	5.957
Total operating costs, $		59,575	59,575	59,575	59,575
Taxable profit, $		15,425	15,425	15,425	15,425
Tax payable at 35%, $		5,399	5,399	5,399	5,399
Cash flow, $	(100,000)	35,027	35,027	35,027	35,027
Discount factor at 15%	*1.000*	*0.870*	*0.756*	*0.658*	*0.572*
Present value, $	(100,000)	30,458	26,485	23,031	20,027
Net present value, $	0				

Note: Numbers in parentheses indicate negative values.

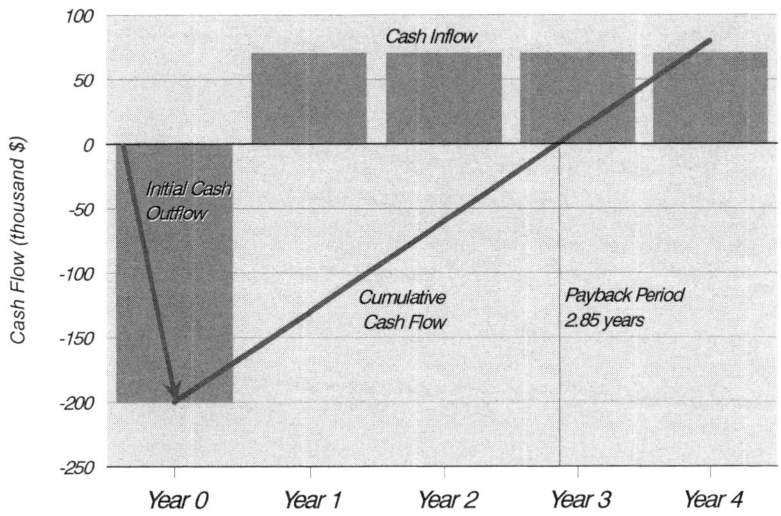

FIGURE 15.5 Traditional payback profile for case A: at-risk capital analysis

In this example, the worst case involves an immediate drop in selling price of 20% and an increase in operating costs by 15%. Since these are externally sourced inputs, they apply to both cases. Internally sourced inputs may not apply equally to both cases—for example, one mining method might be more sensitive to unknown material characteristics than another. The *normal* situation would involve worst-case scenarios that are different for each alternative. Yet even with identical changes in worst-case outside influences, the difference in return on risk capital is substantial between the two cases. Only by

TABLE 15.5 Deterministic worst-case cash flow, case A

	Year				
	0	1	2	3	4
Production		10,000	10,000	10,000	10,000
Unit revenue, $		8.00	8.00	8.00	8.00
Total revenue, $		80,000	80,000	80,000	80,000
Claimable depreciation, $		50,000	50,000	50,000	50,000
Unit operating costs, $		2.202	2.202	2.202	2.202
Total operating costs, $		22,021	22,021	22,021	22,021
Taxable profit, $		7,979	7,979	7,979	7,979
Tax payable at 35%, $		2,792	2,792	2,792	2,792
Cash flow, $	0	55,186	55,186	55,186	55,186
Discount factor at 15%	*1.000*	*0.870*	*0.756*	*0.658*	*0.572*
Present value, $	0	47,988	41,729	36,286	31,553
Net present value, $	157,555				

TABLE 15.6 Deterministic worst-case cash flow, case B

	Year				
	0	1	2	3	4
Production		10,000	10,000	10,000	10,000
Unit revenue, $		8.00	8.00	8.00	8.00
Total revenue, $		80,000	80,000	80,000	80,000
Claimable depreciation, $		25,000	25,000	25,000	25,000
Unit operating costs, $		6.851	6.851	6.851	6.851
Total operating costs, $		68,511	68,511	68,511	68,511
Taxable profit, $		(13,511)	(13,511)	(13,511)	(13,511)
Tax payable at 35%, $		(4,729)	(4,729)	(4,729)	(4,729)
Cash flow, $	0	16,218	16,218	16,218	16,218
Discount factor at 15%	*1.000*	*0.870*	*0.756*	*0.658*	*0.572*
Present value, $	0	14,103	12,263	10,664	9,273
Net present value, $	46,302				

Note: Numbers in parentheses indicate negative values.

subjecting both cases to the same changed circumstance, uninfluenced by idiosyncratic elements, does the real impact of the change become evident.

Tables 15.5 and 15.6 set out the cash flow now applicable. The opportunity cost of capital (the discount rate) has not changed, although this too is not necessarily fixed. If major changes have occurred, then other opportunities may have arisen that now offer greater returns than the 15% value originally applied.

TABLE 15.7 Risk capital under worst-case conditions

Component of Initial Capital	Case A	Case B
Capital not at risk	\$157,555 (78.8%)	\$46,302 (46.3%)
Capital at risk	\$42,445 (21.2%)	\$53,698 (53.7%)

1. The model assumes that the initial capital expenditure has already occurred. The original investment of \$200,000 in case A is now worth \$157,555. Similarly, the original \$100,000 investment in case B is now worth only \$46,302.
2. The initial capital expenditures can still be depreciated from the full amount for tax purposes.
3. The case B worst-case scenario involves tax losses, which in this analysis are assumed to be available to offset taxable profits elsewhere in the company.

If the analysis truly represents the worst case, then the NPV under this worst-case scenario (i.e., this particular component of the original capital) is not at risk. The amount of capital at risk is shown in Table 15.7. The capital that is not at risk has (theoretically, at least) a 100% chance of achieving the hoped-for return. Any amount of this capital can be accommodated in the capital structure of the company, and the effect will be neutral so long as it achieves the same return as the long-term financing cost. Accordingly, this nonrisk component of the capital is "required" to cover only its cost–in this case 10%.

At-Risk Analysis: Return on Risk Capital

The capital that is at risk must be at the center of management focus, and the returns on *this* capital are the ultimate source of growth for the company. Anyone can put their money into nonrisk investments (and retrieve it at no loss), so these types of investment seldom present opportunities for profit. Better projects are the ones that achieve the highest return on the *risk component* of the capital.

In the third step in an at-risk capital dissection, the *original* cash flow is subdivided into two components: a tranche representing the nonrisk element and a tranche representing the risk element. The earliest cash flows constitute the nonrisk tranche, and these are directed at paying back the nonrisk component at the financing cost-of-capital discount rate. The cash flow stream remaining *after this nonrisk return has been achieved* represents the risk tranche, and this is applied to paying back the risk component of the initial capital. The discount rate that equates the present value of the risk tranche to the at-risk capital is the effective return on the risk component of the original capital.

The two cases are set out in Tables 15.8 and 15.9. The presentation in these tables shows a clear difference between the two cases. Case A has a lower amount of its capital at risk and a lower *proportion* of its initial capital at risk.

TABLE 15.8 Risk elements of cash flow for case A

	Year				
	0	1	2	3	4
Cash flow (from 15.3), $	(200,000)	70,053	70,053	70,053	70,053
Cash flow, **nonrisk basis**, $	(157,555)	70,053	70,053	47,883	0
Discount factor at 10%	*1.000*	*0.909*	*0.826*	*0.751*	*0.683*
Present value, $	(157,555)	63,685	57,895	35,975	0
Net present value, $	0				
Cash flow, **risk capital**, $	(42,445)	0	0	22,170	70,053
*Discount factor **at rate of 23.06% to set NPV of risk cash flow to zero***	*1.000*	*0.813*	*0.660*	*0.537*	*0.436*
Present value, $	(42,445)	0	0	11,897	30,548
Net present value, $	0				

Note: Numbers in parentheses indicate negative values.

TABLE 15.9 Risk elements of cash flow for case B

	Year				
	0	1	2	3	4
Cash flow (from 15.4), $	(100,000)	35,027	35,027	35,027	35,027
Cash flow, **nonrisk basis**, $	(46,302)	35,027	17,496	0	0
Discount factor at 10%	*1.000*	*0.909*	*0.826*	*0.751*	*0.683*
Present value, $	(46,302)	31,842	14,460	0	0
Net present value, $	0				
Cash flow, **risk capital**, $	(53,698)	0	17,530	35,027	35,027
*Discount factor **at rate of 16.77% to set NPV of risk cash flow to zero***	*1.000*	*0.856*	*0.733*	*0.628*	*0.538*
Present value, $	(53,698)	0	12,857	22,000	18,841
Net present value, $	0				

Note: Numbers in parentheses indicate negative values.

Effective return on the risk capital is about 23%. Conversely, case B, with a lower overall capital requirement, is clearly more sensitive to the worst-case scenario. It has a higher amount of capital at risk, a much higher proportion of its capital at risk, and consequently a much lower (16.77%) return on its risk capital. Case A should be selected.

GUIDELINES FOR THE AT-RISK CAPITAL APPROACH

Although the analysis in the preceding example study favors the more capital-intensive case, this is not always the result. Typically, more capital-intensive alternatives are also less flexible, with a higher proportion of capital at risk in the worst-case scenario.

The at-risk analysis also signals the appropriate (or at least the *maximum*) debt:equity ratio for the project, and it provides an objective mechanism for subsequent evaluations based only on equity contributions. In this respect, the tabulation is very familiar to analysts involved in the structuring of finance for major projects—the only difference in this instance being the use of the same tools for initial *selection* and *planning* of projects.

Worst-case scenarios are always prepared for major projects; however, without an analytical tool to quantitatively compare alternatives, treatment of those alternatives in the decision process may be quite superficial. The study of worst-case scenarios *as a guide to mine planning* is something that is undertaken only occasionally. Most insiders have difficulty envisaging worst-case scenarios. To be most useful, these scenarios should not just involve remodeling an unchanged mine plan. Changes to the plan are required.

Consider, for example, a case for which unpredicted poor performance from one item of equipment is the major contributor to unsatisfactory *project* performance. The bottleneck introduced by this shortfall in performance reduces total project output by 30%. Since all other capital is sized for 100% performance, the marginal return from incremental investment in equipment to make up for this shortfall will be very large. Unfortunately, the unpredicted shortfall means that (at least from a *finance* perspective) the marginal cost of capital for this incremental investment is also very high. Lenders do not like providing *even more* finance to projects that have yet to prove they can perform. In simple terms, when you, as the owner of a nonperforming mine, go back to the bank to ask for more money, you have no leverage at the bargaining table. Indeed, the risk of overstressing what might be a fragile business relationship with the financiers may be sufficient to inhibit any attempt to do so. The project languishes with substantial inefficiency.

Now consider an alternative scenario. A previously agreed-upon option to call upon additional funds can often be negotiated initially at little cost. There might be a slightly higher interest rate on these funds if they have to be called upon, but this higher rate is a small cost compared to the cost of *not* having the funds available. The rationale is that a firm *knows* that parts of its plan will prove inadequate; it just does not know which *particular* parts. It is more economical to underdesign initially and then institute remedial work than risk overdesign in areas that will prove unnecessary. The at-risk analysis can foreshadow these requirements and demonstrate the value of real options (foreshadowed potential changes to the mine plan) from a risk reduction perspective.

Before project commencement, it may not be possible to predict the occurrence of the worst-case scenario, but at least the capital structure of the project can be put in place ahead of time so that, if there is a problem, shareholder value will not be reduced further by the inability to adjust to it.

CHAPTER 16

Mining Strategy and Knowledge

Throughout the text, references to "knowledge" have pervaded almost every discussion. Explicit *knowledge* about the relative economics associated with the depth and grade of ore was shown in Chapter 2 to be "a valuable guideline for exploration." Chapter 3 suggested that savings in capital are possible if expenditure could be delayed until there was better *knowledge* of the mining conditions. In these examples, knowledge contributes by directly reducing costs or increasing the value of information obtainable for the same cost.

Most references to knowledge are not so easily quantifiable. Chapter 12 examined decision making and queried how *much* knowledge is needed to make a decision and how much risk is associated with the knowledge that individuals do not have and *do not know* that they do not have. This chapter also talked about leaving a legacy of knowledge to subsequent owners–a knowledge externality.

If choices could be made under an assumption of perfect knowledge, then they could be made with mathematical precision. Admittedly this would involve a complex calculation, but if decisions are less than perfect now, is this due to analytical intractability or the imperfection of the knowledge? This final chapter looks at mining strategy from an information and knowledge perspective and recognizes that information (and sometimes *mis*information) is an important tool in formulating strategy.

Example 16.1:

If you have high sunk costs but low cash costs, and your potential competitors have knowledge of this, they might be deterred from entering your market, knowing that you will not easily be put out of business. Alternatively, if your customers have the same knowledge–and if they believe you will continue producing so long as the price exceeds your cash cost–you are vulnerable to exploitation. What should be your strategy? To what extent can you assume that your customers and competitors will act rationally, and/or that in formulating their strategy that they will assume that you will act rationally?

The view of mining strategy set out in this chapter places strong emphasis on these knowledge effects. It suggests that, in mining at least, imperfections in knowledge play a significant role in determining mining strategy and are also a significant contributor to less-then-perfect decision making evidenced throughout the mining world. It follows the general literature on business strategy in emphasizing the role of learning—in both a personal and an organizational sense—and the importance of institutions embodied in corporate culture to harness and capture the value of knowledge within firms. It suggests some promising future directions for mining strategy and enhanced decision making in the mining industry.

STAGES OF STRATEGIC MANAGEMENT

Throughout the whole history of mining, the challenges facing the industry have changed, and the guidelines directing mining project development have had to adjust accordingly. New guidelines must continue to be tested and adopted, and old guidelines discarded. Following Rumelt et al. (1994), Burgelman (1985), and others, five recognized stages in the advance of strategic management of business enterprises are evident:

1. Intuition.
2. Financial planning.
3. Forecast-based planning.
4. Shareholder value focus.
5. Organizational learning.

Each of the stages involves a set of analytical tools to aid decision making, as well as a characteristic or dominant organizational culture. The culture and tools are also identified by the way that the outside environment is treated. The following sections follow this outline and set out some historical background for the first four of these stages, along with the key elements of each stage as they apply to mining-related choices. The fifth stage is addressed in the section of this chapter entitled "Mining Strategy: Where to Go From Here?"

Although the stages of strategic management just outlined represent advancement in chronological terms, subsequent stages enhance rather than supersede preceding stages. Thus, earlier stages remain relevant.

In addition, ideas in strategic management that find expression first at management level also impact lower levels in the corporate hierarchy—if not directly or not immediately, then certainly via changes in corporate culture. For example, mine site personnel (initially at least) might be concerned only with production, whereas the vagaries of commodity pricing remain the preserve of senior management. In due course, commodity price changes *do* impact mine site personnel—through continual change of plans, mine expansions and contractions, nonreplacement of equipment, layoffs, and the like. Site personnel start to concern themselves with product pricing, if not in an analytical way, at least through subtle changes in corporate culture. Thousands of day-to-day decisions will start to be influenced in small ways by the envisaged changes that might have to be made the next time the commodity price changes. Thus, within any one organization, management might be concerning itself with the later stages of strategic management, but the success of an organization might be equally or more dependent on the degree of assimilation of earlier stages throughout the complete workforce.

Stage 1: Mining Management by Intuition

Throughout history, mining has been treated as a different type of business than other businesses. The butcher and the baker may have had to compete for the patronage of customers in the marketplace, but the risks in mining coupled with the uniqueness of skills often insulated mining from many of the competitive pressures faced by consumer-oriented businesses.

In the developed world, this first stage of management—referred to earlier as management by intuition—probably characterized much decision making until the mid-twentieth century. In the less developed world, much decision making today still fits this model—and appropriately so. The key attributes for "success" are, or were, *experience* and *intuition.*

In the environment characterized by this stage of strategic development, managers at mine sites are the key decision makers, since it is the technical aspects of the mining operation that are the dominant concern. Success comes from introducing structure to what is hitherto an unstructured world—be it consistency in vision, refined organizational and reporting skills, or systematic methods of understanding complex orebodies.

Historically, this model has proved quite appropriate, and in many cases it remains so today. For instance, many orebodies have been discovered because of their visible outcrop, and when mined from the outcrop this type of deposit yields immediate cash flows. Sophistication of mining and financing techniques is not called for when almost all development expense can be funded from cash flow. Indeed, mining techniques and mining economics can be learned by trial and error. No real recognition of the impact of the outside world is necessarily called for.

From an economically based decision viewpoint, bigger project developments also present little difficulty if they continue to follow this model—they can be funded from the proceeds of earlier mining. Risk in this environment primarily means *technical* risk—and this is something that mine managers intuitively understand. With new funds sourced from retained earnings, the need to quantify risk elements to outsiders is limited.

Historically, in the developed world, marketing of mine output until the mid-twentieth century also followed this model. For many mineral products, marketing involved selling to the government (gold projects) or to relatively captive and/or regulated local markets (coal, iron ore). Runge (1995) has set out some of the history and culture that underpinned mining decisions from the time of the industrial revolution in the early 1800s and has demonstrated many similarities with the technological revolution being experienced in the late twentieth century.

Is this model relevant today? From the perspective of a mine manager or of anyone who is mainly concerned with technical risk, the model is directly relevant. Experience and intuition remain invaluable (perhaps *the* most valuable) aids to success in a production environment. Limitations revolve around interaction with the outside world—particularly interactions having to do with competition, finance issues (including equity and debt), and risk in a return-on-investment sense.

Stage 2: Financial Planning and Control

In the case of larger and more complex mines, as well as organizations with portfolios of mines, additional management tools are called for. The growth of companies and the availability of more sophisticated technology after World War II were the triggers for these changes in the mining industry. Financial planning tools to aid mining decisions were adopted from elsewhere in the business world.

The most common financial planning aids identified with this stage of management strategy include (in addition to standard sets of accounts) sales, cost, and profit projections; the annual budgeting process; and discounted cash flow analysis of new capital-spending proposals. Almost all of the tools outlined in the first 11 chapters of this book belong to this (second) stage of strategic management and remain relevant today.

Financial planning is the starting point of any strategy for creating value for stockholders. Large enterprises can be subdivided from a financial planning perspective into profit centers or cost centers, and the performance of each component of the business can be understood accordingly.

From a mining industry perspective, three significant shortcomings remain unaddressed if management strategy does not extend beyond this traditional financial planning approach. The limitations exist because

1. The tools basically assume either an unchanging environment or an environment having characteristics that can be estimated through extrapolation of past events. Clearly the outside world today is changing in ways that leave these sorts of projections wanting.
2. The tools focus on efficiency rather than effectiveness. "There is surely nothing quite so useless as doing with great efficiency what should not be done at all" (Drucker 1963).
3. The tools provide little or no assistance for decisions subject to risk (or where risk is different across choices).

These shortcomings were (and still are, in some parts of the mining industry) addressed in a number of ways.

Until the mid-1970s or early 1980s, the assumption of a relatively unchanging environment *was* an appropriate one for much mining. With major growth in steel, aluminum, and electrical demand, new developments in iron ore, coal, and other bulk commodities were frequently underpinned by long-term contracts that provided this consistency. These long-term contracts for project offtake limited the market risk as well as the financing risk. By developing new large mines underwritten by such contracts, mines could proceed on the assumption of supply to a nonchanging outside world—an ideal circumstance for good planning.

The question of efficiency or effectiveness has always been different in the mining industry than in other industries. For example, in manufacturing industries, budgeting processes can easily assume that the future will be just a carbon copy of the past—after all, factory processes do not necessarily change much from year to year. In mining, budgeting has never had the luxury of such a relaxed practice because the orebody to be mined next year can never be assumed the same as the one this year. Exploration provides the mining industry with the kind of renewal and changed management focus that Drucker (1963) sees as vital for success. This partly addresses the efficiency or effectiveness question, but it still leaves unaddressed the issue of how new practices are recognized and introduced, as well as how outdated practices unrelated to orebody changes are discarded.

Financial planning tools to this day do a poor job of understanding risk. Nevertheless, until the mid-1970s this too was seldom an issue of concern in mining. Since the major risks for most mining developments were of a technical or operational nature, and since the boards of most major mining companies consisted mainly of former operating personnel, these higher-level decision makers still had a good intuitive understanding of the technical and operational risks involved in their decisions. Technical risks may not be quantifiable, but as long as the intuition and experience of board-level decision makers are relevant, this inability to quantify risk is not an issue for reliability in decisions.

Stage 3: Forecast-Based Planning

The degree of stability that characterized much mining until the 1970s does not apply to very much mining in the developed world today. Increases in world trade have projected the industry into the forefront of competition and the attendant change. This third stage of strategic management—referred to earlier as the forecast-based planning stage—addresses the changing environment. It explicitly recognizes that the future will not be just an extrapolation of the past, and it focuses management attention on elements of the future that are at least conceptually predictable, as well as on establishing the bounds of possible alternative futures.

New and enhanced analytical tools are associated with this stage of strategic management. Market research is one such tool; at least in the manufacturing industry, it has assumed an important role since the 1960s. Product life cycles, the cyclical pricing and demand for mineral commodities, and the influence of inflation are additional inputs that are recognized in this stage for their increasing importance. These inputs can readily be incorporated in discounted cash flow analysis and in other management decision tools.

Cash flow calculations have also evolved in other ways. Probabilistic sensitivity analysis and simulation, mentioned in Chapter 14, are just two products of this evolution. Since the early 1980s, financial models using spreadsheets and personal computers have taken over the lion's share of economic analysis work. In manufacturing, if not so much in mining, the influence of experience curves on production cost has been recognized and its impact also incorporated into cash flows and competitive analysis. The twin elements of market share and cash costs—discussed already in Chapter 13—became and remain important in management's toolbox of decision aids in this stage of management strategy.

In the resources industries, this phase of strategic management is also home to some of the greatest advances through the development of scenario planning. Many mining companies are now pushing the scenario-planning approach further down the organization rather than leaving it as the exclusive preserve of some executive think tank at the head office. This is a commendable trend.

A scenario is not a prediction. Rather, it is a vehicle for helping people learn. The transition from a management mind-set that does not recognize a changing environment to one that does recognize change is summed up by Schwartz (1991):

> *Often, managers prefer the illusion of certainty to understanding risks and realities. If the forecaster fails in his task, how can the manager be blamed? But in the long run, this denial of uncertainty sets the stage for surprises, shattering the manager's confidence in his or her ability to look ahead. Scenarios allow a manager to say, "I am prepared for whatever happens."* ***It is this ability to act with a knowledgeable sense of risk and reward that separates both the business executive and the wise individual from a bureaucrat or a gambler.*** *(pp. 6–7; emphasis in original)*

Scenarios first emerged after World War II as a method for military planning, but they reached a new dimension in the resources industries in the 1970s and 1980s. Wack (1985) refers to scenario planning as "the gentle art of reperceiving"; the significance of this approach is best illustrated with an example.

Example 16.2:

With few exceptions, mines commence at the shallowest and/or highest-grade sections of the orebody and progress to parts of the orebody that are less economically attractive. Moreover, new discoveries are generally deeper and lower grade than existing mines. Common sense says that the cost of mining, and with it the price of mineral commodities, must rise over time. Indeed, this "fact" seems so intuitively obvious that it goes unquestioned for most practitioners brought up in the physical world of exploration and production.

However, the evidence suggests an entirely different picture. Baumol and Blackman (1993) describe work by themselves and others demonstrating that "the real cost (price) of extraction for a sample of thirteen minerals had declined for all but two (lead and zinc) between 1870 and 1956" and that "the price of fifteen resources for the period 1900 to 1986...until the 'energy crises' of the seventies [showed] negligible upward trend in the real (inflation-adjusted) prices" (pp. 40–41).

A manager relying upon intuition would not perceive this (declining prices) trend. Acting from intuition, he or she would be primarily concerned with managing costs so that they don't increase at a rate faster than the general industry rate of reserve depletion. Yet, as in so many other cases of economics, intuition is misleading. The *effective* stocks of natural resources are continually expanded by the same technological developments that have fueled the extraordinary growth in living standards since the industrial revolution; this growth has, at the same time, fueled the demand for the minerals. Any economic analysis that assumes increased demand in mineral output due to advances in living standards (fueled by technological changes) must also assume declining real prices caused by the same technological changes.

The recognition of this trend is just one counterintuitive outcome that scenario planning can highlight. This type of planning does not predict the future, but it challenges and changes the rules by which future events are judged in the minds of decision makers. With a mind-set alert to a larger range of possible outcomes, decision makers can react more quickly. When changes occur that require plans to be revised, decision makers are not frustrated by an inability to understand what is happening.

Stage 4: Shareholder Value Focus

The forecast-based planning approach goes a long way in addressing the issue of a changing world environment, but it still leaves unaddressed the issue of efficiency versus effectiveness, and it does not explicitly consider risk beyond technical risk. The fourth and fifth stages of strategic management don't explicitly focus on these issues; instead, they direct management attention to them in a much broader way—that is, by focusing on shareholder value.

The shareholder value approach is summed up by Reimann (1988): "The various value-based approaches are all aimed explicitly at the goal of structuring and managing a corporation in a way that will create more value for its shareholders" (p. 10). In short, the focus is on the market price of a corporation's shares—a price that reflects all of the issues already discussed.

Porter (1980; 1985) is identified with much initial work in this stage of strategic management, as well as with pioneering the application of such economics tools as industry analysis and value chain analysis in a business sense. In addition, this stage of strategic management represents the initial application of strategic thinking using the concept of strategy as presented in this book. This concept treats the environment of business *not* as one of passivity and neutrality, but one where other participants—employees, customers, competitors, suppliers—are intelligent and purposeful people whose aims, expected actions, and expected reactions have to be taken into account in arriving at decisions.

Whereas observers such as Porter have developed specific tools for analysis and to guide decision making, others have concentrated on tools to measure management performance under value-focused guidelines. This measurement focus has also resulted in additional guidelines to aid decision making. The at-risk example study (see "Example Study: At-Risk Capital Calculation," p. 235) and the application of DCF tools as set out in Chapter 15 are explicitly focused on this (shareholder value) objective. In this technique, decisions by managers start from the cost-of-capital benchmark, which is directly related to the market value of a firm's debt and equity. Shareholder value grows only when expected or real returns exceed the cost of capital, and projects analyzed under this method first separate out the logical parts of investment projects that do not contribute to growth. As stated in Chapter 15, "Anyone can put their money into nonrisk investments (and retrieve it at no loss), so these types of investments seldom present opportunities for profit. Better projects are the ones that achieve the highest return on the *risk component* of the capital" (p. 239).

Stewart follows the convention strongly echoed throughout this book when he suggests that investors "are only interested in cash, and accounting conventions just muddy the waters. Among those conventions that create the most serious distortions are depreciation and capitalization policies, inventory valuation,...and amortization of goodwill" (quoted in Reimann [1988], p. 12). Stewart (1991) uses a measure of management performance known as economic value added (EVA™), which looks at value (either real or expected) from a shareholder's perspective before and after some (planned or actual) management action. The measure uses the entire capital employed, not just the equity component of the capital.

The focus on shareholder value is well recognized in business, but recognizing the importance of something is not the same as providing the tools to achieve it. Valuations have to be quantifiable, and even the *valuation* for this shareholder perspective is far from straightforward. "This problem of establishing the true value of a business as a going concern is one of the most critical and controversial in applying the value-based planning approach.... The difficulty is that we have no direct and reliable method for assigning a market value to an individual business unit" (Reimann 1988, p. 15). This issue is vital. The success of a whole business quoted on the stock exchange can readily be measured based on the price of shares of stock. Yet how can an individual decision maker contemplating a new mine investment—an investment that is relatively small in comparison to the firm as a whole—bring into the calculation the kinds of indicators that, if followed, will ultimately translate into higher market value?

Fahey and Felton (1988), while not disputing the correctness of focus on value-based planning, consider the measurement of value (before and after a decision is made) to be a significant limitation in wider application of the technique to cases in which the part of the business being valued is separate from the business as a whole.

> *For example, cash-flow forecasts are predicated upon (among other things) sales and revenue forecasts. These forecasts, in turn, are based upon projections of competitors' strategies, technology developments, and customer responses—in short, the industry or competitive context within which rivalry between firms takes place. If any of these projections are badly off the mark, the resultant cash-flow forecasts are likely to be seriously over- or understated. What this means is that value-based planning analysis is only as good—or as bad—as the analysis of the strategic context....Proponents of value-based planning would do well to remember that*

shareholder value analysis is but another element in strategic thinking—not a substitute for it. (p. 4)

The tools and techniques set out in this book have all been presented with this shareholder value objective in mind—aiming to address in a quantitative way many of the values that the value-based planning approach has hitherto addressed only in a conceptual way. The balance of this chapter describes current economic thinking toward extension of this objective and suggests possible mechanisms for implementation in a mining-related sense.

MINING STRATEGY: WHERE TO GO FROM HERE?

The rigors of international competition have caused mining companies the world over to reexamine the way their organizations are run and the key performance indicators they use in judging business success. Phrases such as "shareholder value" and "economic value added" are key parts of this lexicon at senior management level and are starting to enter the consciousness of management elsewhere in mining organizations. The models described in the first 15 chapters of this text broadly follow this value focus.

Shareholder Value

Against this background, it is a useful starting point to reexamine what shareholder value means in quantitative terms, and how this compares with the "value" from NPV calculations and the like as extensively discussed in the first 15 chapters. The following sets out some brief statistics from the 2002 annual report of Rio Tinto, one of the world's leading mining companies.*

Total shareholders' funds	$ 7,462 million
Share market valuation (1,377 million shares at $20 each)	$27,500 million

The owners of Rio Tinto, through their purchases and sales of shares, are valuing their company at the end of 2002 at $27.5 billion, whereas the internal accounts of the company value it at only $7.5 billion.

A company of this size is involved in continual purchases and sales of equipment and even whole mines, so in aggregate the value of individual assets on the books of the company is unlikely to be much different than realizable value in the marketplace. The difference between the share market valuation and the internal accounts is unlikely to be due to errors in asset valuation from an accounting perspective. Clearly the value of Rio Tinto is more than the value of the sum of its parts as measured by standard accounting rules. This begs the question: If it were possible for someone with $7.5 billion to put together a portfolio of assets similar to Rio Tinto, would the share market then value this disparate collection of assets at $27.5 billion?

* Rio Tinto is a dual trading entity comprising Rio Tinto plc of the United Kingdom and Rio Tinto Limited of Australia. Shareholders funds are quoted in company literature in U.S. dollars. The share price in December 2002 of US$20 equates to approximately UK £12.50 or A$35 per share, at the exchange rate of £1 = US$1.6 = A$2.81 used in the company annual report. The number of shares excludes shareholdings by Rio Tinto plc in Rio Tinto Limited.

The share market rates the value of Rio Tinto to be $3^1/2$ times its book value because this is the extra value that Rio Tinto *as an organization* adds. The extra value comes about in many ways. It is in the form of tacit knowledge. It is in the institutionalized procedures that allow hundreds of people to work together. It is in the short-form jargon and culture that is understood by the people in each work environment, pertaining only to that environment or a narrow set of similar work situations. To constitute real extra value, this tacit knowledge cannot be easily replicable; if it were, the way would be clear for someone to buy similar assets and gain the $3^1/2$-times markup that Rio Tinto enjoys. "Valuable resources are those that are superior in use, hard to imitate, difficult to substitute for, and more valuable within the firm than outside.... [V]aluable resources, in order to yield profits to the firm, [have to be] acquired for less than their intrinsic value" (Porter 1994, p. 446).

The important result, however, is that for every $1 of value tied up in identifiable assets, there is more than $2.50 of value in intangible assets. Where should management's efforts be focused? Should efforts be focused on the technical and financial aspects of production and equipment usage for which value is largely quantified by the recognized value (the $7.5 billion) shown on the balance sheet? Or should efforts be focused on expanding the knowledge, learning capability, and efficiency of institutions within the company at large (the added $20 billion of value recognized in the stock market)? This is the challenge for strategic thinkers aiming to add value in mining enterprises: to recognize where this value comes from, when it is likely to be eroded or lost, how to sustain it, and how to create new and unique value-adding institutions and opportunities within the organization.

Dynamic Approach to Mining Strategy

In a world that is relatively consistent from one year to the next, strategic decision making (i.e., formulating the value of choices and making decisions that take into account the actions and likely reactions of other participants) can largely be learned through trial and error. In a dynamic, fast-changing world, the luxury of the trial-and-error approach is not available. There is no laboratory for tests to be carried out and techniques to be proven before implementation, and there is no unambiguous measure at the end to say whether the outcome was the result of the earlier decision. What are the primary elements that underpin good strategic decisions in this environment?

Porter (1994) suggests three promising lines of inquiry toward this objective:

1. The game theory approach.
2. The commitment and uncertainty approach.
3. The resource-based view.

The Game Theory Approach The game theory approach applies specific tools from economics to understand human interaction in ways not previously possible in any quantitative way. Porter (1994) suggests that by "concentrating sequentially on small numbers of variables, the models [of business strategy] fail to capture the simultaneous choices over many variables that characterize most industries" (p. 443). Nevertheless, in mining, many applications can indeed be faithfully represented by using relatively small numbers or sets of variables. For example, in the competitive bidding for supply of raw materials (coal to a base load power station, limestone to a cement mill), there are generally only a small number of viable competitors, each of which has inputs that are generally well known (deposit characteristics, location, etc.). Nalebuff and Brandenburger (1996) have set out examples of the use of game theory in a large number of business applications.

Game theory has also been used to explain and develop models for human interaction. For example, Peters and Waterman (1982) recognized that, among other attributes, the best-run companies could be identified by a distinct openness—a "can do" cooperative approach between employees within the organization. Yet they provided only a limited explanation as to how such an approach came about. Vanberg (1994) has used game theory to understand human interaction in a similar context (though not specifically directed at business applications). Vanberg addresses the following question: "Why do the conventions that enable society to cohere survive, even when it is not in everyone's interests to obey them?" He examines and sets out convincing explanations of why and how the rules and institutions that are the basis of cooperation in society endure. Axelrod (1984) also examines and develops some challenging results from a theoretical perspective using game theory models.

The Commitment and Uncertainty Approach The second promising line of inquiry from Porter (1994) concerns commitment and uncertainty and has direct relevance for the mining industry.

> *The notion here is that strategy is manifested in a relatively few investment decisions, which are hard to reverse and which tend to define choices in other areas of the firm. These commitments must be made under uncertainty.... [T]his approach tends to stress the value of flexibility in dealing with change rather than the capacity to rapidly improve and innovate to nullify or overcome it. (pp. 443–444)*

This approach is directly relevant to strategic choice in the minerals industry, though its importance is not equal across all sectors of the industry. For example, an investment in an underground mine (the majority being in shafts, drives, and access development) contains a much higher proportion of sunk or irreversible cost than a similar amount of capital in an open pit mine. Most examples of strategy in this book derive from literature following this commitment and uncertainty approach.

Example 16.3:

Assume that you own a deposit that can supply important raw materials (limestone and graded aggregate, say) to local customers and that costs of production are sensitive to the scale of production. Assume also that there are other deposits in the general vicinity (owned by others). Customers are spread throughout the region, and the transport distance from individual deposits to individual customers influences the competitiveness significantly. What difference would it make if you adopt a mining method that has high capital (sunk) costs and low operating (cash) costs, compared to a method that has

> low sunk costs? What difference would it make if your competitors and customers knew (or did not know) your cost structure?
>
> If you adopt a capital-intensive mining method that has a high proportion of sunk costs, it may be in your best interest to have this information widely known—at least among potential competitors. A potential competitor who has the capacity to produce at lower overall (average) cost than you, knowing that it is in your best interest to keep producing (even to *increase* production) when prices decline, might be deterred from entering the market. If already in production, a competitor would also be unwise to start a price war because, even if you went bankrupt, the equipment still exists and production from your deposit will not necessarily cease—it will merely be taken up again by the purchaser of your impaired assets. High sunk costs are a deterrent—however, notice that the deterrent value is not actually in the high sunk costs themselves, but in the competitor's *belief* that this is your cost structure. If you actually have high sunk costs but your competitor or potential competitor does not recognize this, or does not recognize its importance, then high sunk costs are not an advantage.
>
> High sunk costs associated with capital-intensiveness are a two-edged sword when it comes to customers. Customers love to see their suppliers making irreversible investments in unique plant that has little value elsewhere and that yields low marginal costs. This knowledge of your cost structure among *customers* is a liability and invites exploitation. Nevertheless, notice the knowledge and informational symmetry here with respect to the competitor situation. High sunk costs are a liability when it comes to customers; again, though, the liability is not necessarily in the high sunk costs themselves, but in the customers' belief that this is your cost structure. If you actually have high sunk costs but your customers do not recognize this, or do not recognize its importance, then high sunk costs are not a liability.

The preceding example illustrates not only the importance of commitment and uncertainty coupled with game theory, but also the importance of the knowledge possessed by and relative sophistication of customers and competitors (both real and potential). It also demonstrates what Porter (1985, p. 212) calls a "good" competitor—in this instance, one who is sufficiently alert both to his or her own cost structure and to your cost structure so as not to take actions that would not be in either party's best interest in the marketplace. Sadly, a lot of mining does not fit this good-competitor model.

There is no doubt that these irrecoverable cost factors have a profound effect on mining investment strategy.

The Resource-Based View The third promising line of inquiry from Porter (1994) concerns the origins of competitive advantage. This approach includes the notion of core competencies and treatments that stress intangible assets. "The argument is that the origins of competitive advantage are valuable resources (or competencies) that firms possess, which are often intangible assets such as skills, reputation, and the like" (Porter 1994, p. 445). There is no doubt that an understanding of a firm's core competencies is essential for rational and consistent decision making, but this does little to aid management in how to develop such competencies in the first place or even to sustain and renew competencies that become dated.

Porter is troubled by much of the debate, particularly the discussion of Prahalad and Hamel (1990), and suggests that the resource-based view risks a circular argument: "Successful firms are successful because they have unique resources. They should nurture these resources to be successful" (Porter 1994, p. 445). Beyond recognizing what is and what isn't a skill within an organization, the core competency argument for maintaining and enhancing competitive advantage therefore rests on discovery of new competencies and on enhancement of existing competencies. For growing an organization, this is a strategy involving *entrepreneurship.* It involves recognizing hitherto unrecognized skills and leveraging existing skills and learning processes in ways that are hard for competitors to emulate. It involves the development of new processes of organizational learning. Some of these issues are addressed in the following section.

NEW TOOLS FOR STRATEGIC MANAGEMENT OF MINING ORGANIZATIONS

Competitive advantage and the maintenance and growth of shareholder value rest firmly on strategies that are unique to the organization. Anything written in a generalized text such as this one—able to be deployed by any mining company—cannot be unique and will necessarily fall short of the mark. Nevertheless, in seeking a generalized strategy for mining-related enterprises, at least four specific directions offer potential: entrepreneurship; capital value and asset management; the value of dispersed and tacit knowledge; and human capital, including the institutions within a firm and the environment in which the firm operates. Much of the potential strategic gain is best understood within a view of the firm that is quite different from traditional, hierarchically structured enterprises. This alternative view of the firm is also addressed in this section.

Entrepreneurship

Entrepreneurship involves recognizing value in some activity or course of action that hitherto has been unrecognized, convincing others of this value, and capturing as much of the value—for oneself or for the benefit of the organization. The quintessential entrepreneurial activity in mining is the discovery of new orebodies, though even this as an example is questionable because most discoveries are now the direct result of logical, systematic, hard work in the areas of research, exploration, and testing.

Entrepreneurship by its nature is less definable *ex ante* than almost any other activity in the business world. The definition of and attributes associated with entrepreneurship cannot be applied after the fact (when the activities are successful); this complicates the notion of entrepreneurship as a meaningful tool for management. Successful businesspeople prefer to attribute their success to entrepreneurial skill and hard work (and not at all the result of luck), whereas unsuccessful business ventures are almost always attributed to bad luck.

> *A theory of entrepreneurial learning (viewed* ex ante*) cannot be a theory of* ex post *successful learning. Success depends upon many factors, including luck (cf. Popper 1976, p47). An adequate theory of learning cannot exclude human fallibility which can manifest itself in entrepreneurial error, losses, and failure. On the contrary, genuine uncertainty, the predictability of future growth of knowledge and the potential for entrepreneurial mistakes and losses must all be emphasized. (Harper 1996, p. 7)*

Kuhn (1970) has documented cases of entrepreneurial scientific discovery and the conditions leading to their acceptance in the scientific community. Barker (1992) has followed the paradigmatic approach of Kuhn and demonstrated application of the scientific concepts to business.

The entrepreneurial approach is similarly seen in Burgelman and Maidique (1988) as the most important next step in advancement of strategic management thinking:

> *At the end of our efforts to describe some of the more complex management processes in large, established firms, we feel even more strongly than at the outset that a theory of corporate entrepreneurship is needed. As once-excellent companies lose their luster and new ones are emerging as bright new stars, it seems clear that simply looking for exemplars of success, whose practices can be readily emulated, is not a workable alternative for serious theory-building efforts....[W]e believe that [the theory of corporate entrepreneurship] will be grounded in increased understanding of the evolutionary processes of organizational learning. (p. 594)*

What is clear is that most advances through successful entrepreneurial activity are recognized, after the fact at least, by something that is then obvious but was not so before. In popular usage, this is the "why didn't I think of that?" phenomenon, or, as promoted in many applications, out-of-the-box thinking. The question could be reformulated to ask what it is that stops these "obvious" alternatives from being recognized beforehand. From an institutional learning perspective, the answer is fairly straightforward. Most organizations are good at what they do best *because* they focus so strongly on this objective. A narrow focus, by definition, excludes consideration of aspects of the situation that initial conceptual filters and time-honored rules of conduct and evaluation deem to be irrelevant to the task at hand. What this means is that the *best* organizations frequently have the poorest ability to embrace new ideas and changes that initially have no direct recognizable relationship to the existing business. The ideas are not consciously or deliberately excluded or suppressed–they are simply not recognized as important.

Senge (1990a; 1990b) has identified circumstances in which core modes of conduct and core value systems inhibit recognition of new ideas (what he calls *core dilemmas*):

> *Management teams typically come unglued when confronted with core dilemmas. A classic example was the way U.S. manufacturers faced the low cost–high quality*

> *choice. For years, most assumed that it was necessary to choose between the two. Not surprisingly, given the short-term pressures perceived by most managements, the prevailing choice was low cost. Firms that chose high quality usually perceived themselves as aiming exclusively for a high quality–high price market niche. The consequences of this perceived either-or choice have been disastrous, even fatal, as U.S. manufacturers have encountered increasing international competition from firms that have chosen to consistently improve quality* ***and*** *cost. (Senge 1990b, p. 18)*

What lessons for mining can be drawn from this example? One parallel with the either-or choice between quality and cost is the risk-return trade-off. In the financial markets—or in any market where information is low cost or free and where there are no unique inputs to the process—there is a distinct relationship between risk and return. The relationship is far from reliable, however, for individual investments and decisions in a mining environment, where information is costly and idiosyncratic and choices are not subject to unrestrained competition because of the presence of unique inputs (the orebody) in the process. Yet many mining investments have been and continue to be made based on the implicit assumptions that low returns somehow represent lower risk and that high returns are necessarily riskier. A reexamination of this core assumption using the model of choice under uncertainty set out in this book promises substantial changes and potential profit over alternatives that continue to follow the either-or approach of risk and return.

A Market-Based View of the Firm

The scope for increased shareholder value addressed in the final three sections of this chapter derives in part from a view of the firm that is distinctly different than for traditional organizations. This section outlines this alternative view of the firm. Increasingly, with extended outsourcing and diversification, large enterprises are asking what it is that distinguishes a firm of a certain size from firms of smaller sizes and, indeed, from business units consisting of individual contractors. Traditionally, in economic theory, the answer to this question rests on the efficiencies of transactions between parts of the business when conducted *within* a firm compared to the same transactions with outsiders.

Gable and Ellig (1993) adopt a different approach. They draw a parallel between the firm as a command-based "economy" and the older Soviet-style command-based economies. The now-identified shortcomings of these totalitarian regimes point to a number of similar shortcomings in traditional command-based business organizations. Some elements of Gable and Ellig's recommended "market-based" approach have already been discussed. The example of internal pricing (see "Internal Pricing," p. 65) is one way for individuals working in their own narrow area of the firm's "economy" to focus only on and maximize the returns from their own area of activity. Halal et al. (1993) describe the application of internal markets to whole organizations. Entrepreneurship, discussed previously, is an activity that is also strongly associated with enterprises and countries that allow freedom for individual experimentation, and these freedoms are seldom evident in command-based countries *or* companies.

Beyond entrepreneurship, the first lesson from this approach that has a counterpart in business has to do with knowledge and coordination. The failure of the Soviet-style economies followed in part because

> *a command-and-control system cannot coordinate the millions of economic decisions needed to produce adequate amounts of consumer goods, even simple ones like bread and shoes. In other words, centralized planning of national economies failed for the same reasons that authoritarian business strategies failed: both approaches overlook the severe limitations to any individual's knowledge.... [M]ost corporations still look much more like centrally planned economies than market systems. (Gable and Ellig 1993, p. 6)*

This *dispersed knowledge* issue is followed up later in this chapter under the heading "Dispersed and Tacit Knowledge."

The second lesson that business firms may draw from the failure of the economies built on a command-and-control structure has to do with capital, or *factors of production*. In the 1920s a fierce argument took place in academic circles concerning the impossibility of economic calculation in such an economy. The argument—advanced by Mises (1920) but (apparently) refuted at the time—held that, although goods of immediately recognizable value could be understood in a command-and-control system, capital goods for which value materializes only over time could not be managed or deployed in any rational way without property rights. With the collapse of these command-and-control economies, the most striking examples of inefficiency are evident in undercapitalization and in the misallocation of capital. After 70 years, Mises was shown to be correct. Unsurprisingly, substantial interest is now focused on many of these original arguments, with potential for valuable application in business enterprises. This issue follows.

Capital Value and Asset Management

This management strategy pointer follows from the recognition described in the preceding section. Command-and-control systems have systemic deficiencies in management of capital because the rewards from better use of capital do not flow to the decision maker or do so only in very indirect ways. There are few industries in which this issue is more important than in the mining industry.

Consider for a start the capital allocation problem when an individual is the owner of the capital and is in complete control of his or her own resource allocation. For a house owner, money and effort spent on extensions and improvements (and money not earned while performing these tasks) represent a loss of immediately evident value. Such a choice in favor of a longer-term objective is pursued only if the envisaged value in the future is greater than the more evident value that is foregone today. If title to the property is secure, the return will be fully realized when the property is sold. The extent to which capital improvements, or any choices extending over time, are appropriately balanced against value realizable now is critically bound up with the security of ownership of the resource (the capital good, or factor of production). Any uncertainty in the eventual return to the decision maker biases the choice to favor the near-term alternative. The Soviet-style command economies failed on this measure because whenever a choice like this one had to be made, the outcome strongly favored the alternative that showed the more immediate benefit. With no ownership of the means of production, incentives favored immediate consumption over more efficient but lengthier processes of production.

Individuals within firms normally can't *own* the assets under their control (the ideal situation from this perspective), so alternative mechanisms must be put in place to replicate the incentives that would apply if the assets were owned. It is the appropriateness of these mechanisms that is in question in this section. Within a firm, security-of-tenure issues abound. Compared to choices that demonstrate immediate returns, the personally realized gains and scope for individual benefits from choices that take longer to prove advantageous are very poorly developed in most organizations. Individuals will balance immediate benefit against future benefit only to the extent that they personally can realize some or all of the benefit. The realization of future payoff is a function of how stable the organization is, the length of time that the individual expects to be in the job, the culture in the organization whereby others expropriate some or all of the benefit from the better decisions, and the turbulence or rate of change in the external environment. The issue is potentially *more* relevant with senior personnel in mining companies who have direct control over the purchase and sale of assets, since many organizations rotate senior personnel into and out of their positions over a 2- or 3-year time frame—a time frame much shorter than the effective life of many mining assets. Many junior employees, on the other hand, often stay employed at the same mine for a long time, and if they stand to gain personally in their job from an action now that will bear fruit only in the future, this is the alternative they are more likely to choose.

The discounted average cost calculation in Chapter 5 (see "Discounted Average Cost," p. 64) and the "Asset Management Considerations" section in Chapter 10 (see p. 160) set out mechanisms for addressing this problem in the case of physical assets like mining equipment. The method involves incentive and remuneration systems built around asset valuation and return on assets using modified DCF tools.

Of more importance are the intertemporal choices associated with orebody exploitation. Everyone who has spent time at a mine site recognizes the perennial trade-off between exploiting an already-developed orebody to make the current-month profit look good—but at the expense of longer-term difficulties that come to light only after the mine manager has moved to a new position. New mechanisms—extending the economically based asset valuation tools in this text to include orebody valuation—offer scope for substantial improvement in strategy through incentive alignment akin to full market-based approaches.

Dispersed and Tacit Knowledge

In past eras, the road to success in business rested on many factors—trade skills, proximity to water or coal, or access to capital, for example. Researchers in economics and management increasingly view the road to success as now being dependent on the recognition, nurturing, and exploitation of knowledge. Hayek (1945) is recognized for his seminal contribution in economics. More recently, Drucker (1993) has posited what he calls a "post-capitalist" society founded on knowledge. Runge (1995) has looked at computer applications in mining from an economics and knowledge perspective.

There are distinctive aspects of the knowledge problem that are particularly relevant to mining. One aspect warrants special attention: the recognition that much knowledge—perhaps the most important knowledge—contained within organizations is inarticulable, or what Polanyi (1983, p. 61) calls "tacit." The quintessential example of tacit knowledge is how to ride a bicycle. While many people know how to ride one, it is very difficult or impossible to articulate this knowledge. A comprehensive corporate manual on bicycle riding might help, but having read such a manual a nonrider would still not be able to ride. Experience is a necessary adjunct to formal learning. "Another often-cited example of tacit knowledge is language. When we learn to talk as children, we also learn a complex set of grammatical rules that very few of us can articulate, but all of us use on a daily basis" (Gable and Ellig 1993, p. 48). Again, certain elements of the process can be learned only through experience.

Within mining enterprises, it is this tacit knowledge that is a vital contributor (perhaps *the* main contributor) to the added value referred to previously in this chapter: "It is in the institutionalized procedures that allow hundreds of people to work together. It is in the short-form jargon and culture that is understood by the people in each work environment, pertaining only to that environment or a narrow set of similar work situations." The example of language provides a clue as to how vital this tacit knowledge is in any industry, because the short-form expressions and jargon associated only with that industry are what encapsulate in efficient ways the ideas and concepts that others inexperienced in that industry cannot comprehend. Surely there are few industries more known for such jargon than the mining industry (the computer industry is the standout exception). What strategies can be developed to enhance the value of this tacit knowledge in a mining context?

The first element of any strategy is to recognize the importance of tacit knowledge and not to erode or destroy it (unless the objective is to deliberately do away with outdated practices). Many mining companies faced with cost pressures have reduced their workforces in a way that this knowledge is lost.

The second element of any strategy is to recognize the nature of the tacit knowledge. Tacit knowledge is less likely to be knowledge of "how to do" something and is more likely to be knowledge of "how to understand" something—the unwritten rules of evaluation. Chapter 3 described the importance of simplified operation and reduced complexity at the start of mining operations, as well as the selection of methods that are consistent with the philosophy of mine owners–key elements that relate directly to issues of tacit knowledge (see "Simplified Operation and Reduced Complexity," p. 21).

Because this knowledge is tacit, it can influence decisions without anyone recognizing its contribution. This presents both an opportunity and a threat, and it suggests that a third element of management strategy involves recognizing where within an organization (e.g., functional areas, individuals) contributions can or should be sought. However, this is asking too much. The knowledge is available, but even many of the potential contributors that have the knowledge don't know that they have it or that it is relevant. The question, to paraphrase Hayek (1946), is "not how we can 'find' the people who know best, but rather what institutional arrangements are necessary in order that the unknown persons who have knowledge specially suited to the task are most likely to be attracted to that task" (p. 95). This challenge, the third element of strategy relating to tacit knowledge, reverses the traditional role of the manager as the person who *allocates* resources to the task. This approach offers enormous promise for radical reform in the way tacit knowledge within organizations is capitalized upon.

Human Capital

Some of the (added) value recognized in the share market and enjoyed by *all* major companies comes directly from the people who work in the organization. A manager who is paid a salary of $100,000 per year may add $200,000 of value to the company each year. This does not mean that the manager is underpaid–his or her most attractive *alternative* employment might offer only $70,000 per year. If the manager's knowledge is unique, the value added is a function of the person and the environment in which his or her services are put to use. There is no definitive way to value individual contributions in what is effectively *team* production.

In a world that is changing slowly, knowledge, skills, and equipment remain useful for a long time. In a world that is changing more quickly, both physical equipment and human skills (human capital, or knowledge) have a shorter economic life. Skills that once were valuable in the marketplace are no longer wanted. Engineers who are experts at working a slide rule can no longer command a premium in the mine-planning marketplace.

Drucker (1993, p. 186) highlights that "[k]nowledge formation is...already the largest investment in every developed country.... [S]urely, the return which a country or a company gets on knowledge must increasingly be a determining factor in its competitiveness." Alertness to the importance of change is the starting point in addressing this problem. From an asset management perspective, human capital is by far the biggest resource in most major mining companies. Institutional procedures to maintain this capital in a constant state of effectiveness are a vital element in any strategy for growth and enhancement of shareholder value. Responsibility rests both with organizations and with individuals.

In 1776 Adam Smith published a monumental volume on *The Wealth of Nations* (the title of the book). His description of mining was sobering:

> *Of all those expensive and uncertain projects, however, which bring bankruptcy upon the greater part of the people who engage in them, there is none perhaps more perfectly ruinous than the search after new...mines....Projects of mining, instead of replacing the capital employed in them,...commonly absorb both capital and profit. (Smith 1976 [1776], p. 562)*

More than 200 years later, there remain many mining projects that continue to "absorb both capital and profit." Yet this characterization of mining is hardly a robust one, because resources would not continue to be directed into mining after this time unless there was an expectation in aggregate that returns will materialize. What is it that originally led such an insightful observer as Smith to reach this conclusion and that (apparently) continues to lure investors into mining–only some of whom achieve satisfactory returns?

The answer lies at least in part with ideas first formulated by Smith himself. In marveling at the relative coordination in society, the most efficient aspects of which seemed to work with no conscious or deliberate plan, Smith introduced his concept of the "invisible hand." An actor "pursuing his own interest... frequently promotes that of the society more effectually than when he really intends to promote it" (p. 456).

Within a mining enterprise, profits or losses are certainly a function of management skill and the analytical tools that are brought to bear. However, they are also an outcome of how well the tacit knowledge is harnessed through coordinating mechanisms within the organization akin to the Smithian invisible hand in society. The successes and failures of the 1980s and 1990s suggest that the latter effect is more important than commonly realized and that the companies that have neglected the latter are poorer for the experience. A strategy aimed at building an environment where knowledge can be effectively put to work represents perhaps the greatest promise for value adding available in the changing world of mining today.

APPENDIX A

Financial Tables

This appendix presents tables of values for use in calculations of functions described in the main text of this book (Tables A.1, A.2, A.3, and A.4).

TABLE A.1 Present Value Factors—Present Value (PV) of $1 to Be Received After *n* Years at Interest Rate *i*. PV = $1/(1 + i)^n$

	Rate																			
Year	1%	2%	3%	4%	5%	6%	7%	8%	9%	10%	11%	12%	13%	14%	15%	16%	17%	20%	25%	30%
1	0.9901	0.9804	0.9709	0.9615	0.9524	0.9434	0.9346	0.9259	0.9174	0.9091	0.9009	0.8929	0.8850	0.8772	0.8696	0.8621	0.8547	0.8333	0.8000	0.7692
2	0.9803	0.9612	0.9426	0.9246	0.9070	0.8900	0.8734	0.8573	0.8417	0.8264	0.8116	0.7972	0.7831	0.7695	0.7561	0.7432	0.7305	0.6944	0.6400	0.5917
3	0.9706	0.9423	0.9151	0.8890	0.8638	0.8396	0.8163	0.7938	0.7722	0.7513	0.7312	0.7118	0.6931	0.6750	0.6575	0.6407	0.6244	0.5787	0.5120	0.4552
4	0.9610	0.9238	0.8885	0.8548	0.8227	0.7921	0.7629	0.7350	0.7084	0.6830	0.6587	0.6355	0.6133	0.5921	0.5718	0.5523	0.5337	0.4823	0.4096	0.3501
5	0.9515	0.9057	0.8626	0.8219	0.7835	0.7473	0.7130	0.6806	0.6499	0.6209	0.5935	0.5674	0.5428	0.5194	0.4972	0.4761	0.4561	0.4019	0.3277	0.2693
6	0.9420	0.8880	0.8375	0.7903	0.7462	0.7050	0.6663	0.6302	0.5963	0.5645	0.5346	0.5066	0.4803	0.4556	0.4323	0.4104	0.3898	0.3349	0.2621	0.2072
7	0.9327	0.8706	0.8131	0.7599	0.7107	0.6651	0.6227	0.5835	0.5470	0.5132	0.4817	0.4523	0.4251	0.3996	0.3759	0.3538	0.3332	0.2791	0.2097	0.1594
8	0.9235	0.8535	0.7894	0.7307	0.6768	0.6274	0.5820	0.5403	0.5019	0.4665	0.4339	0.4039	0.3762	0.3506	0.3269	0.3050	0.2848	0.2326	0.1678	0.1226
9	0.9143	0.8368	0.7664	0.7026	0.6446	0.5919	0.5439	0.5002	0.4604	0.4241	0.3909	0.3606	0.3329	0.3075	0.2843	0.2630	0.2434	0.1938	0.1342	0.0943
10	0.9053	0.8203	0.7441	0.6756	0.6139	0.5584	0.5083	0.4632	0.4224	0.3855	0.3522	0.3220	0.2946	0.2697	0.2472	0.2267	0.2080	0.1615	0.1074	0.0725
11	0.8963	0.8043	0.7224	0.6496	0.5847	0.5268	0.4751	0.4289	0.3875	0.3505	0.3173	0.2875	0.2607	0.2366	0.2149	0.1954	0.1778	0.1346	0.0859	0.0558
12	0.8874	0.7885	0.7014	0.6246	0.5568	0.4970	0.4440	0.3971	0.3555	0.3186	0.2858	0.2567	0.2307	0.2076	0.1869	0.1685	0.1520	0.1122	0.0687	0.0429
13	0.8787	0.7730	0.6810	0.6006	0.5303	0.4688	0.4150	0.3677	0.3262	0.2897	0.2575	0.2292	0.2042	0.1821	0.1625	0.1452	0.1299	0.0935	0.0550	0.0330
14	0.8700	0.7579	0.6611	0.5775	0.5051	0.4423	0.3878	0.3405	0.2992	0.2633	0.2320	0.2046	0.1807	0.1597	0.1413	0.1252	0.1110	0.0779	0.0440	0.0254
15	0.8613	0.7430	0.6419	0.5553	0.4810	0.4173	0.3624	0.3152	0.2745	0.2394	0.2090	0.1827	0.1599	0.1401	0.1229	0.1079	0.0949	0.0649	0.0352	0.0195
16	0.8528	0.7284	0.6232	0.5339	0.4581	0.3936	0.3387	0.2919	0.2519	0.2176	0.1883	0.1631	0.1415	0.1229	0.1069	0.0930	0.0811	0.0541	0.0281	0.0150
17	0.8444	0.7142	0.6050	0.5134	0.4363	0.3714	0.3166	0.2703	0.2311	0.1978	0.1696	0.1456	0.1252	0.1078	0.0929	0.0802	0.0693	0.0451	0.0225	0.0116
18	0.8360	0.7002	0.5874	0.4936	0.4155	0.3503	0.2959	0.2502	0.2120	0.1799	0.1528	0.1300	0.1108	0.0946	0.0808	0.0691	0.0592	0.0376	0.0180	0.0089
19	0.8277	0.6864	0.5703	0.4746	0.3957	0.3305	0.2765	0.2317	0.1945	0.1635	0.1377	0.1161	0.0981	0.0829	0.0703	0.0596	0.0506	0.0313	0.0144	0.0068
20	0.8195	0.6730	0.5537	0.4564	0.3769	0.3118	0.2584	0.2145	0.1784	0.1486	0.1240	0.1037	0.0868	0.0728	0.0611	0.0514	0.0433	0.0261	0.0115	0.0053

TABLE A.2 Future Value Factors—Future Value (FV) of $1 Invested Today for *n* Years at Interest Rate *i*. FV = $(1 + i)^n$

	Rate																		
Year	1%	2%	3%	4%	5%	6%	7%	8%	9%	10%	11%	12%	13%	14%	15%	16%	17%	20%	25%	30%
1	1.0100	1.0200	1.0300	1.0400	1.0500	1.0600	1.0700	1.0800	1.0900	1.1000	1.1100	1.1200	1.1300	1.1400	1.1500	1.1600	1.1700	1.2000	1.2500	1.3000
2	1.0201	1.0404	1.0609	1.0816	1.1025	1.1236	1.1449	1.1664	1.1881	1.2100	1.2321	1.2544	1.2769	1.2996	1.3225	1.3456	1.3689	1.4400	1.5625	1.6900
3	1.0303	1.0612	1.0927	1.1249	1.1576	1.1910	1.2250	1.2597	1.2950	1.3310	1.3676	1.4049	1.4429	1.4815	1.5209	1.5609	1.6016	1.7280	1.9531	2.1970
4	1.0406	1.0824	1.1255	1.1699	1.2155	1.2625	1.3108	1.3605	1.4116	1.4641	1.5181	1.5735	1.6305	1.6890	1.7490	1.8106	1.8739	2.0736	2.4414	2.8561
5	1.0510	1.1041	1.1593	1.2167	1.2763	1.3382	1.4026	1.4693	1.5386	1.6105	1.6851	1.7623	1.8424	1.9254	2.0114	2.1003	2.1924	2.4883	3.0518	3.7129
6	1.0615	1.1262	1.1941	1.2653	1.3401	1.4185	1.5007	1.5869	1.6771	1.7716	1.8704	1.9738	2.0820	2.1950	2.3131	2.4364	2.5652	2.9860	3.8147	4.8268
7	1.0721	1.1487	1.2299	1.3159	1.4071	1.5036	1.6058	1.7138	1.8280	1.9487	2.0762	2.2107	2.3526	2.5023	2.6600	2.8262	3.0012	3.5832	4.7684	6.2749
8	1.0829	1.1717	1.2668	1.3686	1.4775	1.5938	1.7182	1.8509	1.9926	2.1436	2.3045	2.4760	2.6584	2.8526	3.0590	3.2784	3.5115	4.2998	5.9605	8.1573
9	1.0937	1.1951	1.3048	1.4233	1.5513	1.6895	1.8385	1.9990	2.1719	2.3579	2.5580	2.7731	3.0040	3.2519	3.5179	3.8030	4.1084	5.1598	7.4506	10.6045
10	1.1046	1.2190	1.3439	1.4802	1.6289	1.7908	1.9672	2.1589	2.3674	2.5937	2.8394	3.1058	3.3946	3.7072	4.0456	4.4114	4.8068	6.1917	9.3132	13.7858
11	1.1157	1.2434	1.3842	1.5395	1.7103	1.8983	2.1049	2.3316	2.5804	2.8531	3.1518	3.4785	3.8359	4.2262	4.6524	5.1173	5.6240	7.4301	11.6415	17.9216
12	1.1268	1.2682	1.4258	1.6010	1.7959	2.0122	2.2522	2.5182	2.8127	3.1384	3.4985	3.8960	4.3345	4.8179	5.3503	5.9360	6.5801	8.9161	14.5519	23.2981
13	1.1381	1.2936	1.4685	1.6651	1.8856	2.1329	2.4098	2.7196	3.0658	3.4523	3.8833	4.3635	4.8980	5.4924	6.1528	6.8858	7.6987	10.6993	18.1899	30.2875
14	1.1495	1.3195	1.5126	1.7317	1.9799	2.2609	2.5785	2.9372	3.3417	3.7975	4.3104	4.8871	5.5348	6.2613	7.0757	7.9875	9.0075	12.8392	22.7374	39.3738
15	1.1610	1.3459	1.5580	1.8009	2.0789	2.3966	2.7590	3.1722	3.6425	4.1772	4.7846	5.4736	6.2543	7.1379	8.1371	9.2655	10.5387	15.4070	28.4217	51.1859
16	1.1726	1.3728	1.6047	1.8730	2.1829	2.5404	2.9522	3.4259	3.9703	4.5950	5.3109	6.1304	7.0673	8.1372	9.3576	10.7480	12.3303	18.4884	35.5271	66.5417
17	1.1843	1.4002	1.6528	1.9479	2.2920	2.6928	3.1588	3.7000	4.3276	5.0545	5.8951	6.8660	7.9861	9.2765	10.7613	12.4677	14.4265	22.1861	44.4089	86.5042
18	1.1961	1.4282	1.7024	2.0258	2.4066	2.8543	3.3799	3.9960	4.7171	5.5599	6.5436	7.6900	9.0243	10.5752	12.3755	14.4625	16.8790	26.6233	55.5112	112.455
19	1.2081	1.4568	1.7535	2.1068	2.5270	3.0256	3.6165	4.3157	5.1417	6.1159	7.2633	8.6128	10.1974	12.0557	14.2318	16.7765	19.7484	31.9480	69.3889	146.192
20	1.2202	1.4859	1.8061	2.1911	2.6533	3.2071	3.8697	4.6610	5.6044	6.7275	8.0623	9.6463	11.5231	13.7435	16.3665	19.4608	23.1056	38.3376	86.7362	190.050

TABLE A.3 Capital Recovery Factors for Interest Rate i and Number of Years n. Capital recovery factor = $i/\{1 - [1/(1 + i)^n]\}$

Year	Rate 1%	2%	3%	4%	5%	6%	7%	8%	9%	10%	11%	12%	13%	14%	15%	16%	17%	20%	25%	30%
1	1.0100	1.0200	1.0300	1.0400	1.0500	1.0600	1.0700	1.0800	1.0900	1.1000	1.1100	1.1200	1.1300	1.1400	1.1500	1.1600	1.1700	1.2000	1.2500	1.3000
2	0.5075	0.5150	0.5226	0.5302	0.5378	0.5454	0.5531	0.5608	0.5685	0.5762	0.5839	0.5917	0.5995	0.6073	0.6151	0.6230	0.6308	0.6545	0.6944	0.7348
3	0.3400	0.3468	0.3535	0.3603	0.3672	0.3741	0.3811	0.3880	0.3951	0.4021	0.4092	0.4163	0.4235	0.4307	0.4380	0.4453	0.4526	0.4747	0.5123	0.5506
4	0.2563	0.2626	0.2690	0.2755	0.2820	0.2886	0.2952	0.3019	0.3087	0.3155	0.3223	0.3292	0.3362	0.3432	0.3503	0.3574	0.3645	0.3863	0.4234	0.4616
5	0.2060	0.2122	0.2184	0.2246	0.2310	0.2374	0.2439	0.2505	0.2571	0.2638	0.2706	0.2774	0.2843	0.2913	0.2983	0.3054	0.3126	0.3344	0.3718	0.4106
6	0.1725	0.1785	0.1846	0.1908	0.1970	0.2034	0.2098	0.2163	0.2229	0.2296	0.2364	0.2432	0.2502	0.2572	0.2642	0.2714	0.2786	0.3007	0.3388	0.3784
7	0.1486	0.1545	0.1605	0.1666	0.1728	0.1791	0.1856	0.1921	0.1987	0.2054	0.2122	0.2191	0.2261	0.2332	0.2404	0.2476	0.2549	0.2774	0.3163	0.3569
8	0.1307	0.1365	0.1425	0.1485	0.1547	0.1610	0.1675	0.1740	0.1807	0.1874	0.1943	0.2013	0.2084	0.2156	0.2229	0.2302	0.2377	0.2606	0.3004	0.3419
9	0.1167	0.1225	0.1284	0.1345	0.1407	0.1470	0.1535	0.1601	0.1668	0.1736	0.1806	0.1877	0.1949	0.2022	0.2096	0.2171	0.2247	0.2481	0.2888	0.3312
10	0.1056	0.1113	0.1172	0.1233	0.1295	0.1359	0.1424	0.1490	0.1558	0.1627	0.1698	0.1770	0.1843	0.1917	0.1993	0.2069	0.2147	0.2385	0.2801	0.3235
11	0.0965	0.1022	0.1081	0.1141	0.1204	0.1268	0.1334	0.1401	0.1469	0.1540	0.1611	0.1684	0.1758	0.1834	0.1911	0.1989	0.2068	0.2311	0.2735	0.3177
12	0.0888	0.0946	0.1005	0.1066	0.1128	0.1193	0.1259	0.1327	0.1397	0.1468	0.1540	0.1614	0.1690	0.1767	0.1845	0.1924	0.2005	0.2253	0.2684	0.3135
13	0.0824	0.0881	0.0940	0.1001	0.1065	0.1130	0.1197	0.1265	0.1336	0.1408	0.1482	0.1557	0.1634	0.1712	0.1791	0.1872	0.1954	0.2206	0.2645	0.3102
14	0.0769	0.0826	0.0885	0.0947	0.1010	0.1076	0.1143	0.1213	0.1284	0.1357	0.1432	0.1509	0.1587	0.1666	0.1747	0.1829	0.1912	0.2169	0.2615	0.3078
15	0.0721	0.0778	0.0838	0.0899	0.0963	0.1030	0.1098	0.1168	0.1241	0.1315	0.1391	0.1468	0.1547	0.1628	0.1710	0.1794	0.1878	0.2139	0.2591	0.3050
16	0.0679	0.0737	0.0796	0.0858	0.0923	0.0990	0.1059	0.1130	0.1203	0.1278	0.1355	0.1434	0.1514	0.1596	0.1679	0.1764	0.1850	0.2114	0.2572	0.3046
17	0.0643	0.0700	0.0760	0.0822	0.0887	0.0954	0.1024	0.1096	0.1170	0.1247	0.1325	0.1405	0.1486	0.1569	0.1654	0.1740	0.1827	0.2094	0.2558	0.3035
18	0.0610	0.0667	0.0727	0.0790	0.0855	0.0924	0.0994	0.1067	0.1142	0.1219	0.1298	0.1379	0.1462	0.1546	0.1632	0.1719	0.1807	0.2078	0.2546	0.3027
19	0.0581	0.0638	0.0698	0.0761	0.0827	0.0896	0.0968	0.1041	0.1117	0.1195	0.1276	0.1358	0.1441	0.1527	0.1613	0.1701	0.1791	0.2065	0.2537	0.3021
20	0.0554	0.0612	0.0672	0.0736	0.0802	0.0872	0.0944	0.1019	0.1095	0.1175	0.1256	0.1339	0.1424	0.1510	0.1598	0.1687	0.1777	0.2054	0.2529	0.3016

TABLE A.4 Sinking Fund Factors for Interest Rate i and Number of Years n. Sinking fund factor = $i/[(1 + i)^n - 1]$

	Rate																			
Year	1%	2%	3%	4%	5%	6%	7%	8%	9%	10%	11%	12%	13%	14%	15%	16%	17%	20%	25%	30%
1	1.0000	1.0000	1.0000	1.0000	1.0000	1.0000	1.0000	1.0000	1.0000	1.0000	1.0000	1.0000	1.0000	1.0000	1.0000	1.0000	1.0000	1.0000	1.0000	1.0000
2	0.4975	0.4950	0.4926	0.4902	0.4878	0.4854	0.4831	0.4808	0.4785	0.4762	0.4739	0.4717	0.4695	0.4673	0.4651	0.4630	0.4608	0.4545	0.4444	0.4348
3	0.3300	0.3268	0.3235	0.3203	0.3172	0.3141	0.3111	0.3080	0.3051	0.3021	0.2992	0.2963	0.2935	0.2907	0.2880	0.2853	0.2826	0.2747	0.2623	0.2506
4	0.2463	0.2426	0.2390	0.2355	0.2320	0.2286	0.2252	0.2219	0.2187	0.2155	0.2123	0.2092	0.2062	0.2032	0.2003	0.1974	0.1945	0.1863	0.1734	0.1616
5	0.1960	0.1922	0.1884	0.1846	0.1810	0.1774	0.1739	0.1705	0.1671	0.1638	0.1606	0.1574	0.1543	0.1513	0.1483	0.1454	0.1426	0.1344	0.1218	0.1106
6	0.1625	0.1585	0.1546	0.1508	0.1470	0.1434	0.1398	0.1363	0.1329	0.1296	0.1264	0.1232	0.1202	0.1172	0.1142	0.1114	0.1086	0.1007	0.0888	0.0784
7	0.1386	0.1345	0.1305	0.1266	0.1228	0.1191	0.1156	0.1121	0.1087	0.1054	0.1022	0.0991	0.0961	0.0932	0.0904	0.0876	0.0849	0.0774	0.0663	0.0569
8	0.1207	0.1165	0.1125	0.1085	0.1047	0.1010	0.0975	0.0940	0.0907	0.0874	0.0843	0.0813	0.0784	0.0756	0.0729	0.0702	0.0677	0.0606	0.0504	0.0419
9	0.1067	0.1025	0.0984	0.0945	0.0907	0.0870	0.0835	0.0801	0.0768	0.0736	0.0706	0.0677	0.0649	0.0622	0.0596	0.0571	0.0547	0.0481	0.0388	0.0312
10	0.0956	0.0913	0.0872	0.0833	0.0795	0.0759	0.0724	0.0690	0.0658	0.0627	0.0598	0.0570	0.0543	0.0517	0.0493	0.0469	0.0447	0.0385	0.0301	0.0235
11	0.0865	0.0822	0.0781	0.0741	0.0704	0.0668	0.0634	0.0601	0.0569	0.0540	0.0511	0.0484	0.0458	0.0434	0.0411	0.0389	0.0368	0.0311	0.0235	0.0177
12	0.0788	0.0746	0.0705	0.0666	0.0628	0.0593	0.0559	0.0527	0.0497	0.0468	0.0440	0.0414	0.0390	0.0367	0.0345	0.0324	0.0305	0.0253	0.0184	0.0135
13	0.0724	0.0681	0.0640	0.0601	0.0565	0.0530	0.0497	0.0465	0.0436	0.0408	0.0382	0.0357	0.0334	0.0312	0.0291	0.0272	0.0254	0.0206	0.0145	0.0102
14	0.0669	0.0626	0.0585	0.0547	0.0510	0.0476	0.0443	0.0413	0.0384	0.0357	0.0332	0.0309	0.0287	0.0266	0.0247	0.0229	0.0212	0.0169	0.0115	0.0078
15	0.0621	0.0578	0.0538	0.0499	0.0463	0.0430	0.0398	0.0368	0.0341	0.0315	0.0291	0.0268	0.0247	0.0228	0.0210	0.0194	0.0178	0.0139	0.0091	0.0060
16	0.0579	0.0537	0.0496	0.0458	0.0423	0.0390	0.0359	0.0330	0.0303	0.0278	0.0255	0.0234	0.0214	0.0196	0.0179	0.0164	0.0150	0.0114	0.0072	0.0046
17	0.0543	0.0500	0.0460	0.0422	0.0387	0.0354	0.0324	0.0296	0.0270	0.0247	0.0225	0.0205	0.0186	0.0169	0.0154	0.0140	0.0127	0.0094	0.0058	0.0035
18	0.0510	0.0467	0.0427	0.0390	0.0355	0.0324	0.0294	0.0267	0.0242	0.0219	0.0198	0.0179	0.0162	0.0146	0.0132	0.0119	0.0107	0.0078	0.0046	0.0027
19	0.0481	0.0438	0.0398	0.0361	0.0327	0.0296	0.0268	0.0241	0.0217	0.0195	0.0176	0.0158	0.0141	0.0127	0.0113	0.0101	0.0091	0.0065	0.0037	0.0021
20	0.0454	0.0412	0.0372	0.0336	0.0302	0.0272	0.0244	0.0219	0.0195	0.0175	0.0156	0.0139	0.0124	0.0110	0.0098	0.0087	0.0077	0.0054	0.0029	0.0016

Glossary

accounting rate of return – Project earnings, after taxes and depreciation, divided by the book value of the investment. Average accounting returns do not discount for the timing of earnings and hence are only of limited use for comparison of alternatives where cash flows differ on an annual basis.

average accounting return – *See* accounting rate of return.

capital – The value, assessed in the present, of a cash flow stream (or anything of value) envisaged in the future. The value is attached to certain goods (capital goods) that—combined with resources, institutional structures, and a plan within some market framework—represent the ingredients in bringing this future to realization.
See also capital goods.

capital goods – Tools, machinery, equipment, and physical facilities normally used in the technological transformation of raw material ingredients into a finished good.

capital-intensive – Pertaining to investments requiring expenditure of resources now, rather than later; in the expectation of a higher return or to avoid even higher expenditure later. Usually this means spending (more) capital at the start of a mine to achieve low or lower operating costs throughout the mine life. The benefits of a capital-intensive approach could also be in the form of lower risk or lower marginal production costs (allowing easier expandability).

cash costs – Costs of production per ton (or over some time period) based only on money actually paid out. Cash costs exclude, for example, depreciation of equipment, depletion of reserves, and capital expenditures. Interest payments on loans are cash payments, but since they are independent of production they are frequently treated separately from *operating* cash costs. Tax payments are also cash costs but are also treated separately in most evaluations.

cash flow – The difference between cash inflow (receipts, or revenue) and cash outflow (actual cash costs, including capital expenditure and tax at the time it is paid). Cash flows are normally grouped by time period in which they are expected to appear. *Note:* Tax paid is a real cash item. Depreciation is not a real cash item—it enters the analysis only as part of the tax calculation.

compounding – Process of reinvesting each interest payment to earn more interest. If an amount of money today (present value, or PV) is compounded at an interest rate (i) over a number of years (n), the future value (FV) will be $FV = PV \times (1 + i)^n$, where $(1 + i)^n$ is the compound factor.

compound interest – Reinvestment of each interest payment on money invested, to earn more interest.

constant dollar evaluation – An evaluation calculating the return *over and above* expected inflation. Costs are escalated, revenues are escalated, and the entire cash flow is de-escalated at the expected inflation rate first, prior to the "normal" discounting being undertaken. A simple constant dollar evaluation just ignores escalation and inflation and yields a result that is approximately correct, assuming all costs and revenues escalate at the same rate. The more precise constant dollar evaluation using escalation and de-escalation accounts for the fact that depreciation, *for tax purposes,* is based on historical values. In an environment of high inflation, an evaluation that *does not* escalate and de-escalate the cash flow first will underestimate the "real" tax payable.

deflation – The opposite of inflation. Deflation is the term normally used when prices go down (i.e., the value of money goes up).
See also inflation.

depreciation – A noncash expense representing the proportion of the cost of plant or equipment that is charged against earnings in the accounting period. The objective is to fairly allocate the degree of "wearing out" of the equipment to the production in the period. Taxation authorities also allow depreciation as a legitimate expense (even though there is no actual expenditure incurred). Depreciation *for tax purposes* may or may not fairly represent the true "wearing out" of the capital goods in question.

discounted average cost – The end result of a form of discounted cash flow analysis. The discounted average cost is the price per unit of production that you would have to pay for an independent operator with the same investment criteria as yourself to undertake the production.

discounted cash flow (DCF) – A table of cash flows, usually presented on a year-by-year basis, wherein future cash flows are multiplied by discount factors to obtain present value.

discounted cash flow rate of return (DCFROR) – Term that, for most applications, can be used interchangeably with the terms *internal rate of return* (IRR) or *return on investment* (ROI). Since all calculations of return imply discounting to a constant point in time, the simpler alternative terms are preferred.

See also internal rate of return; return on investment.

discount factor – Factor that, when multiplied by a future value, yields the equivalent present value.

See also discounting.

discounting – The opposite of compounding. A future value (FV) is discounted to the present value (PV) at a discount rate (i) over a number of years (n) by the formula FV = PV$(1 + i)^{-n}$, where $(1 + i)^{-n}$ is the discount factor.

discount rate – The rate at which a firm values future events compared to present events. It is usually thought of as the extent to which future cash flows must be discounted to yield a value representing what those cash flows would be worth if they were received or paid out today. The discount rate applied to a particular project must be higher than the interest rate incurred by the company on funds borrowed for the same project—because loan repayments precede shareholder dividend payments and therefore are lower risk.

DCF – *See* discounted cash flow.

DCFROR – *See* discounted cash flow rate of return.

EBIT – Earnings before interest and taxation.

EBDIT – Earnings before depreciation, interest, and taxation.

economic analysis – Any analysis that applies economic criteria to the technical or production calculations. In the case of whole projects, what is commonly termed an economic analysis normally implies estimation of capital and operating costs of production as well as revenue from sale of products.

See also financial analysis.

economics – Traditionally the science of the (optimum) allocation of scarce means to attain certain ends. In this context, economics relates to the optimal allocation of limited resources (e.g., personnel skills, money, time) to obtain the greatest return (e.g., highest return on investment, maximum net present value, lowest risk).

effective annual interest rate – A more accurate expression of an interest rate value that has been expressed in simplified terms. Quoted rates of interest (on a bank loan, for example) are often simplified for the benefit of presentation and calculation, and the effective rate may be different than the rate superficially presented. For example, a quoted "annual" interest rate of 12%, calculated at 3% per quarter, is an effective rate of $(1.03)^4 = 1.1255$, or 12.55%.

end-of-year convention – A standard for accounting and tabulating cash flow data that presents all cash flows as occurring at the end of the year in which they actually occur. Capital expenditures that have to be in place at the start of any year are normally tabulated at the end of the previous year.

equity – Investments in the form of ownership titles—usually shares of stock. The equity proportion of the capital in a project is the amount left over after satisfaction of all outstanding debts and other obligations. Equity investments are distinguished from investments in loans, bonds, or other forms of debt, which represent claims that must be met and fully satisfied before any claims, dividends, or other distribution to the owners or shareholders.

escalation – Changes in the price of a specific item over time. Commonly, long-term contracts for earthmoving have escalation clauses wherein the price is adjusted by formula whenever there is an increase in some basic commodity (e.g., fuel oil price).

expected return – Average of possible returns weighted by their probability of occurrence. If possible returns are characterized by a normal distribution, the expected return would be the mean of the distribution.

factor of production – A human service or material good that can be used to contribute to the success of a process of production; a constituent element of any production process. Examples are labor, fuel, and capital goods. Factors of production can be classified as to (1) human (labor) or nonhuman (material) factors or (2) original or produced factors. The term *factor of production* may also refer to the time factor (paint and concrete take *time* to cure/produce).

financial analysis – Comprehensive economic analysis including analysis of (1) the risk in a project, (2) the form of financing, (3) the operating and tax structure, and (4) the effect the project might have on the owners' whole corporate profitability and cash flow position.

inflation – In popular (nonscientific) usage, the change in price of a "basket" or group of items over time. Inflation is popularly considered to be the change in the consumer price index (CPI), which is a weighted index of prices of common commodities purchased by a typical household. Inflation indices are also published for many other baskets of commodities (e.g., index of manufactured goods).

Inflation is the term normally used when prices go up (i.e., the value of money goes down). Inflation depreciates the value of money over time, since in an inflationary environment the same amount of cash will not purchase the same amount of goods in the future.
See also deflation.

income – The returns that come in during some period—usually after subtraction of the direct costs of production. The term is insufficiently well enough defined to stand on its own and should be used in conjunction with more descriptive terms, e.g., net operating income (operating revenues minus operating expenses) or taxable income (for operating income less allowable tax deductions).

interest – The price paid for the use of (someone else's) money. Interest is commonly thought of as having two components: a time preference component, which effectively represents the difference between the present values of present and future goods (due to impatience and opportunity, for example); and an entrepreneurial component, which represents the uncertainty element as to whether the money is likely to be repaid or not. Hence, a dollar today is more valuable than a dollar (likely to be) received in a year's time.

internal rate of return (IRR) – The discount rate that, when applied to all of a project's cash flows (including the initial investments) equates the net present value to zero. It is effectively the same as the return on investment. Return on investment implies an initial investment (cash outflow) and subsequent returns (cash inflow). Many projects require initial investments spread over several years, with further investments (for plant upgrading, equipment replacement, etc.) throughout the life, coupled with variable cash inflows and outflows. The IRR can be calculated internally (i.e., independent of the company's cost of capital and other factors), but a decision to proceed with a project or not would still require a calculation of the discount rate as a basis for comparison with alternative uses of the company's resources.
See also return on investment; net present value.

IRR – *See* internal rate of return.

loss – The opposite of profit. Loss occurs when the money returned from an activity is less than the money put into the activity. This term is insufficiently defined to use on its own. Companies may make accounting losses or tax losses in certain years of a project yet still be achieving a satisfactory return on investment.

marginal cost, marginal revenue – The change in total cost or total revenue of producing an extra unit of output. Profits are maximized when output is expanded to the point where the *marginal* cost of producing the *extra* unit of production just equates to the *marginal* revenue of the product sold. In determining the marginal cost, whether some factors of production (e.g., capital goods) are included or not will depend on the time frame being considered.

money – The most commonly used medium or media of exchange in a market society; a community's most marketable economic good, which people seek primarily for the purposes of later exchanging units of it for the goods and services they prefer.

net present value (NPV) – The sum total of the amounts (whether positive or negative) from a series of cash flows (for cash spent or received at various times in the future) discounted to the present value in a discounted cash flow tabulation. The discount rate applied may be the company's financing cost for the project or (frequently) a rate (greater than the financing cost) that the company considers appropriate for the risk, resources involved, and alternative opportunities available.

nominal dollar evaluation (or accounting) – A process that uses actual values—costs and sales price—expected on the day the cost or revenue is paid/received, assuming inflation and other predicted price trends. The nominal dollar evaluation parallels the actual accounting that is undertaken in day-to-day operations in the mine. This form of evaluation is necessary during feasibility studies for calculation of debt-servicing requirements and for tax accounting purposes.

nominal interest rate – The quoted interest rate (e.g., on a bank loan).

NPV – *See* net present value.

offtake – The sale of product from a mine or other production process. In fully competitive markets, offtake is synonymous with demand, sincc in these markets quantities bear direct relationships with price. In markets characterized by idiosyncrasies (such as many of the markets supplied by mining enterprises), the relationship between price and quantity (offered or demanded) may take many years to adjust, and during the adjustment process a company may not be able to sell its products even at reduced prices.

operating hour (for equipment) – Any hour when the motor of a machine is running. The machine need not be doing productive work during this time (it might be deadheading, or waiting in a queue, for example).

opportunity cost – The value that would otherwise have materialized from the highest-valued alternative scenario of comparable risk that has to be foregone to undertake the project in question. In economics, the "cost" of *anything* is always the opportunity cost because that truly represents what must be given up to pursue the planned activity.

payback – The time that it takes before the money returned from a project equates to the money initially put into the project.

present value – A measure of value (expressed in dollar terms) accounting for the fact that present goods are more valuable than future goods because the present goods are available here and now and the future goods are not. The estimate of the value of future goods (received then), when multiplied by the discount rate, yields the value of the future goods as if they were available now—the present value.
See also discount rate.

profit – The difference between the money returned from an activity and the money put into the activity. Making a profit is the fundamental objective of most business activity. In most mining enterprises, profit is expressed in terms of the after-tax return on investment in the enterprise. The accounting "profit" refers to the money remaining at the end of each time period. It is calculated by taking the revenue earned or received and deducting operating expenses, taxes, and allowances for depreciation of equipment.

real interest rate – An interest rate that reflects the price paid for use of money *above* the amount someone would have to pay just to keep up with inflation. It is obtained by adjusting the nominal interest rate for the rate of inflation. For example, if the nominal interest rate on a personal loan is 20% and inflation is 10%, then the real interest rate is (1.20/1.10) = 1.091 = 9.1%.

return on capital employed (ROCE) – A rate similar to the return on investment but used for operating mines or projects that have a mixture of debt and equity funding. For example, a company might have all of its equipment leased, so its own investment is zero and its return on investment infinite. This is clearly an inappropriate indicator of how well the company's assets are being deployed. ROCE is a more reliable measure (at least for comparing one project performance with another) because it effectively puts all projects on a 100% equity basis (excludes finance risk).

return on investment (ROI) – The discount rate such that the cash (in)flows throughout the project, when discounted to the present, just equate to the initial investment (cash outflow) at the start of the project. The return on investment can be calculated independent of the company's cost of capital and other factors, but a decision to proceed or not would still require a calculation of the discount rate as a basis for comparison. In most applications, the terms *return on investment* (ROI), *internal rate of return* (IRR) and *discounted cash flow rate of return* (DCFROR) can be used interchangeably. The term *return on investment* is preferred where there is just one initial outlay (the investment) followed by subsequent cash inflows. The term *internal rate of return* is preferred where additional investments (after the initial investment) are made throughout the life of the project.
See also internal rate of return.

revenue – Money received (or invoiced) for the sale of product. Cash flow analyses are concerned with money received. Accounting procedures record revenue at the time of invoice.

risk analysis – The use of certain techniques aimed at assessing the overall project risk given uncertainty characteristics of the component costs and revenues. The process involves changing several input parameters simultaneously and randomly but according to their probability of occurrence. After repeated simulations, a statistical distribution of probable results is obtained.

ROCE – *See* return on capital employed.

ROFE (return on funds employed) – *See* return on capital employed.

ROI – *See* return on investment.

salvage value – The expected value (after the costs of selling have been deducted) realized upon disposal of a fixed asset at the end of its useful life.

sensitivity analysis – A process that measures the effects of specified variations (plus or minus) in project parameters on investment criteria. The simplest and most common form of sensitivity analysis examines the effect of one input variable at a time and assumes all others remain constant. Sensitivity analysis does not in itself assess the *risk* of an investment subject to changes in a certain variable. To measure risk, some additional assessment is necessary incorporating the *probability* of the change occurring.

sunk costs – The nonrecoverable part of a fixed cost. Sunk costs usually refer to one-time investments that must be made in entering a market and that have no residual value if the firm exits the market. For a business already in a market (i.e., such that sunk costs are already incurred), sunk costs are not included in cash flow analysis unless they somehow contribute to the calculation of tax payable; however, they may be included in the company accounts (financial accounting) if they relate to activity in subsequent time periods.

taxable profit – Earnings calculated according to a set of government-determined rules that establish the prima facie sum subject to taxation. For example, entertainment expenses may be an integral part of a business but may not be allowable in the government formula for business expenses.

taxation – Money paid to government authorities based on taxable profit earned. Note that, in determining the tax payable, the government usually stipulates a formula for *its* definition of a firm's profit, which may or may not be a reliable indicator of actual profit.

time value of money – Conceptual and analytical device whereby cash flows occurring at different times throughout the life of a project are to be expressed in terms of their equivalent value at some fixed point in time (usually the start of the project).
See also present value; discount rate.

tranche – Part of a single financing or cash flow stream that is structured into different maturities (dates when cash flows occur), different principal amounts, or (sometimes) different currencies.

weighted average cost of capital – Expected return on a portfolio of all the firm's (existing) securities. This value is used as the starting point for assessing new capital investment.

working capital – Funds (cash) that are necessary in a business to cover the running costs caused by the time differences between expenditures incurred to produce the goods and revenue received from the sale of the goods. Spare parts held in the warehouse represent an example of such a cost.

References

Abel, A.B. 1983. Optimal Investment Under Uncertainty. *American Economic Review,* 73:228–233.

Abel, A.B., and J.C. Eberly. 1995. Optimal Investment With Costly Reversibility. NBER Working Paper 5091. Cambridge, Mass.: National Bureau of Economic Research.

Abel, A.B., A.K. Dixit, J.C. Eberly, and R.S. Pindyck. 1995. Options: The Value of Capital, and Investment. NBER Working Paper 5227. Cambridge, Mass.: National Bureau of Economic Research.

Akerlof, G.A. 1984. The Market for "Lemons": Quality Uncertainty and the Market Mechanism. In *An Economic Theorist's Book of Tales.* Cambridge: Cambridge University Press.

Alchian, A.A. 1963. Reliability of Progress Curves in Airframe Production. *Econometrica,* 31(4):679–693.

——. 1968. Cost. *International Encyclopaedia of the Social Sciences,* 3: 404–415. New York: Crowell Collier and MacMillan, Inc.

Alford, C.G., and J. Whittle. 1986. Application of Lerchs-Grossman Pit Optimization to the Design of Open Pit Mines. In *Proceedings, Large Open Pit Mining Conference, Newman, W.A.* Melbourne, Australia: Australasian Institute of Mining and Metallurgy.

Arrow, K.J. 1958. Utilities, Attitudes, Choices: A Review Note. *Operations Research,* 5:765–774. Reprinted in Arrow (1984).

——. 1984. Individual Choice Under Certainty and Uncertainty. In *Collected Papers of Kenneth J. Arrow.* Cambridge, Mass.: Harvard University Press.

Axelrod, R. 1984. *The Evolution of Cooperation.* New York: Basic Books.

Barker, J.A. 1992. *Paradigms: The Business of Discovering the Future.* New York: HarperCollins.

Baumol, W.J., and S.A.B. Blackman. 1993. Natural Resources. In *The Fortune Encyclopaedia of Economics.* Edited by D.R. Henderson. New York: Warner Books.

Baumol, W.J., J.C. Panzar, and R.D. Willig. 1982. *Contestable Markets and the Theory of Industry Structure.* San Diego, Calif.: Harcourt Brace Jovanovich.

Brealey, R.A., and S.C. Myers. 2003. *Principles of Corporate Finance.* New York: McGraw-Hill.

Brennan, M.J., and E.S. Schwartz. 1985. Evaluating Natural Resource Investments. *Journal of Business,* 58(2):135–137 (University of Chicago Press).

Buchanan, J.M. 1969. *Cost and Choice: An Inquiry in Economic Theory.* Chicago, Ill.: University of Chicago Press.

——. 1981. Introduction: L.S.E. Cost Theory in Retrospect. In *L.S.E. Essays on Cost.* Edited by J.M. Buchanan and G.F. Thirlby. New York: New York University Press.

Burgelman, R.A. 1985. Managing Corporate Entrepreneurship: New Structures for Implementing Technological Innovation. *Technology in Society,* 7:91–103.

Burgelman, R.A., and M.A. Maidique. 1988. *Strategic Management of Technology and Innovation.* Homewood, Ill.: Irwin.

Carlton, D.W., and J.M. Perloff. 1994. *Modern Industrial Organization.* New York: HarperCollins.

Crowsen, P. 1991. *Minerals Handbook, 1990–91.* London: MacMillan and Co.

Dixit, A.K., and B. Nalebuff. 1991. *Thinking Strategically.* New York: W.W. Norton & Company.

Dixit, A.K., and R.S. Pindyck. 1994. *Investment Under Uncertainty.* Princeton, N.J.: Princeton University Press.

Drucker, P.F. 1963. Managing for Business Effectiveness. *Harvard Business Review,* May-June(63303) (Harvard University Press).

——. 1970. *Technology, Management, and Society.* London: Heinemann.

——. 1993. *Post-Capitalist Society.* New York: HarperCollins.

Fahey, L., and S. Felton. 1988. The Bottom Line on Value-Based Planning. *Planning Review,* 16(1):4–5.

Ferderer, J.P. 1993. The Impact of Uncertainty on Aggregate Investment Spending: An Empirical Analysis. *Journal of Money, Credit, and Banking,* 25:30–48.

Gable, W., and J. Ellig. 1993. *Introduction to Market-Based Management.* Fairfax, Va: Center for Market Processes.

Gentry, D.W., and T.J. O'Neil. 1984. *Mine Investment Analysis.* New York: Society of Mining Engineers of the American Institute of Mining, Metallurgical, and Petroleum Engineers.

Halal, W.E., A. Geranmayeh, and J. Proudehnad. 1993. *Internal Markets: Bringing the Power of Free Enterprise Inside Your Organization.* New York: John Wiley & Sons.

Harper, D.A. 1996. *Entrepreneurship and the Market Process.* London: Routledge.

Hayek, F.A. 1945. Use of Knowledge in Society. *American Economic Review,* 35(4):519–530 (American Economic Association).

——. 1946. The Meaning of Competition. In *Individualism and Economic Order.* Edited by F.A. Hayek. Chicago, Ill.: University of Chicago Press.

Hurn, A.S., and R.E. Wright. 1994. Geology or Economics? Testing Models of Irreversible Investment Using North Sea Oil Data. *Economic Journal,* 104: 363–371.

Knight, F.H. 1985 (1921). *Risk, Uncertainty, and Profit.* Chicago, Ill.: University of Chicago Press.

Kuhn, T.S. 1970. *The Structure of Scientific Revolutions.* Revised edition. Chicago, Ill.: University of Chicago Press.

Lachmann, L.M. 1978. *Capital and Its Structure.* Kansas City, Kan.: Sheed Andrews and McMeel, Inc.

Lane, K.F. 1991. *The Economic Definition of Ore.* London: Mining Journal Books Limited.

Leahy, J.V., and T.M. Whited. 1995. The Effect of Uncertainty on Investment: Some Stylized Facts. NBER Working Paper 4986. Cambridge, Mass.: National Bureau of Economic Research.

Lerchs, H., and I.F. Grossman. 1965. Optimum Design of Open Pit Mines. *Canadian Mining and Metallurgy Bulletin,* 58(633):17–24.

Lewis, W.A. 1948. *Overhead Costs.* New York: Rinehart and Company, Inc.

Lintner, J. 1965. The Valuation of Risk Assets and the Selection of Risky Investments in Stock Portfolios and Capital Budgets. *Review of Economics and Statistics,* 47(February).

Locke, J. 1995 (1693). *An Essay Concerning Human Understanding.* New York: Prometheus Books.

Love, B. 1997. Strategic Decision Making: Its Application by Mining Corporations. *Management,* July:24–26 (Sydney: Australian Institute of Management).

Lucas, R.E., and E.C. Prescott. 1971. Investment Under Uncertainty. *Econometrica,* 39(5).

Markowitz, H.M. 1952. Portfolio Selection. *Journal of Finance,* 7(March): 77–91.

McDonald, R., and D. Siegel. 1986. The Value of Waiting to Invest. *Quarterly Journal of Economics,* 1986:707–727.

Mises, L. von. 1920. Economic Calculation in the Socialist Commonwealth. Originally written in German, translated and reprinted in Hayek, F.A., ed. 1935. *Collectivist Economic Planning.* London: Routledge.

——. 1966. *Human Action: A Treatise on Economics.* Chicago, Ill.: Henry Regnery Company.

Morrison, C.J. 1993. Investment in Capital Assets and Economic Performance: The U.S. Chemicals and Primary-Metals Industries in Transition. *Journal of Business and Economic Statistics,* 11(1):45–60 (American Statistical Association).

Nalebuff, B.J., and A.H. Brandenburger. 1996. *Co-opetition.* London: HarperCollins.

Peters, T.J., and R.H. Waterman. 1982. *In Search of Excellence: Lessons From America's Best Run Companies.* New York: Warner Books.

Pindyck, R.S. 1988. Irreversible Investment, Capacity Choice, and the Value of the Firm. *American Economic Review,* December:969–985.

——. 1993. The Present Value Model of Rational Commodity Pricing. *The Economic Journal,* 103:511–530 (Cambridge: Royal Economic Society).

Pindyck, R.S., and A. Solimano. 1993. Economic Instability and Aggregate Investment. In *NBER Macroeconomics Annual, 1993.* Cambridge, Mass.: National Bureau of Economic Research and MIT Press.

Polanyi, M. 1983. *The Tacit Dimension.* New York: Doubleday.

Popper, K.R. 1976. *Unended Quest: An Intellectual Autobiography.* London: Fontana/Collins.

Porter, M.E. 1980. *Competitive Strategy: Techniques for Analyzing Industries and Competitors.* New York: The Free Press.

——. 1985. *Competitive Advantage: Creating and Sustaining Superior Performance.* New York: The Free Press, Macmillan, Inc.

——. 1994. Toward a Dynamic Theory of Strategy. In *Fundamental Issues in Strategy.* Edited by R.P. Rumelt, D.E. Schendel, and D.J. Teece. Boston, Mass.: Harvard Business School Press.

Prahalad, C.K., and G. Hamel. 1990. The Core Competence of the Corporation. *Harvard Business Review,* May-June:79–91.

Quirin, G.D. 1967. *The Capital Expenditure Decision.* Homewood, Ill.: Richard D. Irwin, Inc.

Reimann, B.C. 1988. Managing for the Shareholders: An Overview of Value-Based Planning. *Planning Review,* 16(1):10–22.

Rio Tinto Limited. 2002. *2002 Annual Report and Financial Statements.* London: Rio Tinto Limited.

Roditis, Y.S. 1993. Beyond Open Pit Optimization Planning, Scheduling and Sensitivity Analysis. In *Proceedings, SME Annual Meeting, Reno, Nevada, 1993.* Littleton, Colo.: Society for Mining, Metallurgy, and Exploration.

Rothschild, M.L. 1992. Bionomics: The Inevitability of Capitalism. London: Futura Publications, Macdonald and Company (Publishers).

RTZ-CRA. 1996. *1996 Annual Report to Shareholders.* London: RTZ-CRA Limited.

Rumelt, R.P., D.E. Schendel, and D.J. Teece. 1994. *Fundamental Issues in Strategy.* Boston, Mass.: Harvard Business School Press.

Runge, I.C. 1988. Strip Ratio: An Outdated Indicator of Economic Value. In *Proceedings, International Symposium on Mine Planning and Equipment Selection.* Calgary, Canada: A. A. Balkema.

——. 1990. Assessing Risk and Capital Investment Criteria in Mining Projects. In *Proceedings, Mincost '90.* Sydney, Australia: Australasian Institute of Mining and Metallurgy.

——. 1992. Maintaining and Improving the Economics of Open Pit Mining. Conference Theme Address. In *Proceedings, Third Large Open Pit Mining Conference.* Melbourne, Australia: Australasian Institute of Mining and Metallurgy.

——. 1994. Uncertainty and Risk in Mineral Valuation: A User's Perspective. In *Proceedings, Mineral Valuation Methodology (VALMIN '94)* Conference. Sydney, Australia: Australasian Institute of Mining and Metallurgy.

——. 1995. Economics, Technological Change, and the Knowledge Problem. In *Proceedings, 25th APCOM Conference: Application of Computers in the Mineral Industry.* Melbourne, Australia: Australasian Institute of Mining and Metallurgy.

——. 2000. *Capital and Uncertainty.* Cheltenham, U.K.: Edward Elgar.

Schwartz, P. 1991. *The Art of the Long View.* New York: Doubleday.

Senge, P.M. 1990a. *The Fifth Discipline: The Art and Practice of the Learning Organization.* New York: Random House.

——. 1990b. The Leader's New Work: Building Learning Organizations. *Sloan Management Review,* Fall:7–23.

Shackle, G.L.S. 1983. The Bounds of Unknowledge. In *Beyond Positive Economics.* Edited by J. Wiseman. London: Macmillan.

Sharpe, W.F. 1964. Capital Asset Prices: A Theory of Market Equilibrium Under Conditions of Risk. *Journal of Finance,* September: 19.

Shepherd, W.G. 1984. "Contestability" vs. Competition. *American Economic Review,* 74:572–587.

Smith, A. 1976 (1776). *An Inquiry Into the Nature and Causes of the Wealth of Nations.* Edited by R.H. Campbell and A.S. Skinner. Oxford, England: Oxford University Press.

Stewart, G.B. 1991. *The Quest for Value: The EVA™ Management Guide.* New York: HarperCollins.

Stigler, G.J. 1961. The Economics of Information. *Journal of Political Economy,* 69(3):213–225.

Strydom, P.D.F. 1986. The Economics of Information: A Subjectivist View. In *Subjectivism, Intelligibility and Economic Understanding.* Edited by I. Kirzner. New York: New York University Press.

Summers, L.H. 1987a. Corporate Capital Budgeting Practices and the Effects of Tax Policies on Investment. In *Taxes and Capital Formation.* Edited by M. Feldstein. Chicago, Ill.: University of Chicago Press.

——. 1987b. Investment Incentives and the Discounting of Depreciation Allowances. In *The Effects of Taxation on Capital Accumulation.* Edited by M. Feldstein. Chicago, Ill.: University of Chicago Press.

Trigeorgis, L. 1996. *Real Options: Managerial Flexibility and Strategy in Resource Allocation.* Cambridge, Mass.: MIT Press.

Vanberg, V.J. 1994. *Rules and Choice in Economics.* London: Routledge.

Wack, P. 1985. Scenarios: Uncharted Waters Ahead. *Harvard Business Review,* September/October:73–89 (Harvard University Press). First part of a two-part series entitled "The Gentle Art of Reperceiving."

Whittle, J. 1988. Beyond Optimization in Open Pit Design. In *Proceedings of the First Canadian Conference on Computer Applications in the Mineral Industry*. Rotterdam, Netherlands: A.A. Balkema.

Whittle, J., and L.I. Rozman. 1991. Open Pit Design in the 90s. In *Mining Industry Optimisation Conference*. Melbourne, Australia: Australasian Institute of Mining and Metallurgy.

Index

Note: *f* indicates figure, *t* indicates table, *n* indicates footnote.

Biography

Ian C. Runge, a corporate advisor on capital investment strategy, is executive director of Capital Strategy Pty Ltd. and founder/nonexecutive chairman of Runge Limited, a mining consulting, software, and training firm based in Australia with offices around the world.

Runge's expertise extends to mining technology, finance, and investment as it relates to gold, industrial minerals, and coal. Recognized as a world authority in mining economics and mine design, Runge lectures and travels extensively. A fourth-generation mining practitioner, he holds a master's degree in mining engineering as well as a doctorate in economics and has authored two books and more than 20 papers on topics related to mining technology, finance, and investment.

Runge is past chairman of the Australasian Institute of Mining and Metallurgy (Southern QLD) and frequently chairs or organizes conferences and other industry forums.

Spreadsheet and Financial Modeling Software

Please complete the form below and fax, e-mail, or return it to *Runge Mining* to receive

- Spreadsheet files of the examples included in the text.
- A licensed copy of XERAS/Lite—a limited version of *Runge Limited*'s powerful XERAS™ financial modeling software. XERAS/Lite has most of the features of XERAS™ but is limited to five years of analysis and basic mining activities. XERAS/Lite is an educational tool, designed for ease of use; it builds on many of the fundamental economic concepts discussed in the text.
- Updates and notification of additional materials.

Prof./Dr./Mr./Mrs./Ms. Name:

Position in Company/Company:

Address:

City: State: ZIP/Postal Code:

Country:

Phone (Work): Fax:

E-mail Address, Home Page (URL):

Fields of Interest Relating to Mining Economics and Strategy:

POST, FAX, OR E-MAIL TO:

Runge Limited
G.P.O. Box 2774
Brisbane, Qld.
Australia 4001

Phone: (07) 3100-7200
Fax: (07) 3100-7297
E-mail: irunge@runge.com
http://www.runge.com

☐ I am the original purchaser of the book. Please forward the latest version of XERAS/Lite and spreadsheet models at no charge.

☐ Please forward the latest version of XERAS/Lite and spreadsheet models. I enclose US$ _____, being for (____) copies at US$40 per copy (or equivalent).

Copies of software (in CD-ROM format) and other materials will be forwarded from the nearest *Runge Limited* office. For prices in other currencies; evaluation copies for educational and training institutions, and corporate training departments; and alternate addresses for forwarding please contact the office above. E-mail is the fastest and most efficient way to contact *Runge Limited.* If e-mailing, use "Mining Economics and Strategy" in the "Subject" field.